# Laser welding

## A practical guide

# Laser welding

## A practical guide

Christopher Dawes, CEng

ABINGTON PUBLISHING

Woodhead Publishing Ltd in association with The Welding Institute
Cambridge England

Published by Abington Publishing
Woodhead Publishing Limited, Abington Hall, Granta Park,
Great Abington, Cambridge CB21 6AH, England
www.woodheadpublishing.com

First published 1992, Abington Publishing
Reprinted 2008



British Library Cataloguing in Publication Data
A catalogue record for this book is available from the British Library.

ISBN 978-1-85573-034-2

The publisher's policy is to use permanent paper from mills that operate a sustainable forestry policy, and which has been manufactured from pulp which is processed using acid-free and elementary chlorine-free practices. Furthermore, the publisher ensures that the text paper and cover board used have met acceptable environmental accreditation standards.

# Contents

# Preface

Laser welding is one of the newest welding techniques available to industry. It is not merely an alternative technique, but a technique which offers fabrication opportunities hitherto difficult or impossible by other welding methods. Laser welding, however, is quite different to other fusion welding techniques from both equipment and process operation standpoints. Moreover, the name laser for some people conjures up a science fiction image which can cause some reticence in accepting the laser as a welding tool.

Although several books and hundreds of scientific papers have been written about lasers and laser processing, there is little information dedicated to the practice of applying laser welding. Most users initially rely on the machine supplier for advice for the selection of equipment and welding conditions and thereafter build on their own experience with respect to the overall operation. To change this situation this book has been written to ease the way for those wishing to understand and apply the practical aspects of laser welding.

This book is intended to introduce the laser as a welding tool and then to provide a practical user's guide to the selection, implementation and overall operation of laser welding. The first three chapters aim to give a basic understanding and general feel for the process, by describing how the laser works, the characteristics of laser welding which distinguish it from other welding techniques and a diverse range of industrial applications where the laser is used to advantage. The remaining chapters provide technical and practical information for those considering or currently using laser welding. The many aspects covered include materials which can be welded; selecting welding parameters; joint configurations; essential equipment; quality control and process monitoring; laser safety; and installing and operating a laser. A general glossary of terminology related to lasers and welding is also provided.

I would like to convey my thanks to my wife Pat who both encouraged me and patiently typed the manuscript. I would also like to thank my colleague Steve Riches for giving his time to reading the manuscript and making constructive comments.

Some of the technical data presented in this book was kindly supplied by the following:

Conrad Banas (USA)
Daimler-Benz (Germany)
Edward Metzbower (USA)
Lumonics Limited (UK)
Philips (The Netherlands)
Rofin Sinar (USA)
Thyssen Stahl (Germany)
Trumpf (UK)
II-VI Inc (USA)
TWI (UK)
Utilase Systems Inc (USA)
Vickers Systems Limited (UK)

*Christopher Dawes*

Chapter 1

# Light and lasers

This chapter describes the release of energy from light waves; the difference between laser light and sunlight; the basic principle of how a laser works; the different types of lasers used for welding; and harnessing the laser beam to achieve a high power density.

## Light

The natural light by which we see is also the essential energy source by which our planet survives; it controls plant growth, the climate and the weather. The light rays from the sun, which travel at about 300 000 km/s and take approximately 8.5 minutes to arrive on the earth's surface, impart their energy according to the sun's different light wave lengths and the substances on which they fall. Naturally selected light wave lengths when absorbed by the molecules of certain substances excite the molecules causing them to vibrate and generate heat. A steel plate beside, or even placed behind a plain glass plate in the full sunlight, on a hot summer's day, will absorb more energy from the light waves than the glass plate. The steel will become so hot that it will be uncomfortable to touch, whereas the glass will not be significantly warmer than the air through which the light rays passed. Glass absorbs less heat because it is transparent to most of the light wave lengths which make up the sun's rays. The results of this comparison, however, can be changed if a specific wave length is artificially produced. Steel, for example, is partially transparent to certain X-rays. These artificial rays are the same type as light, but of much shorter wave length.

The ability of the different light wave lengths to give up their energy in the form of heat when absorbed by different substances, coupled with the fact that light can be transmitted long distances and then be optically focused to a small spot and cause a massive increase in its power density ($W/mm^2$), has led scientists and engineers to develop special light making machines called lasers.

## Lasers

Lasers produce a collimated and coherent beam of light (coherent: waves of one wave length all in phase). This light is quite different from the incoherent light of the sun, Fig. 1.1, which radiates in all directions from its source. The almost parallel, single wave length, light rays which make up the collimated laser beam have a considerably higher power density and can be focused to a much smaller spot size than the randomly radiated rays of the sun. Consequently, a much more efficient power density is achieved with a laser.

The word LASER is an acronym, it stands for: (L) light (A) amplification (S) stimulated by the (E) emission of (R) radiation, and refers to the way in which the light is generated. The basic principle of how a laser works is presented in the following paragraph. Those seeking a more detailed scientific explanation are recommended to read the books by Koechner[1] and Duley.[2]

All lasers are optical amplifiers which work by pumping (exciting) an active medium placed between two mirrors, one of which is partially transmitting, Fig. 1.2. The active medium is a collection of specially selected atoms, molecules or ions which can be in a gas, liquid or solid form and which will lase, i.e. emit radiation as light waves (referred to as photons) when excited by the pumping action. Pumping of liquids and solids is achieved by flooding them with light from a flash lamp and gases are pumped by applying an electrical discharge.

The term photon is used instead of light wave when describing the production of laser light, because the photon carries with it a precisely defined amount of energy in relation to its wave length.

Whatever the active medium consists of: atoms, molecules or ions there are billions of them and they absorb energy when pumped, which they hold for a very short but random life time. When their life time expires they give up their energy in the form of a photon and return to their former state until pumped again. The release of photons in this manner is called spontaneous emission. The photons released travel in all directions in relation to the optical axis of the laser, Fig. 1.3. If a photon collides with another energised atom, etc, it causes it to release its photon prematurely and the two photons will travel along in phase until the next collision, thus building a stream of photons of increasing density, Fig. 1.4. This action of releasing a photon prematurely is called stimulated emission. Photons which do not travel parallel to the optical axis of the laser are quickly lost from the system. Those which do travel parallel to the axis have their path length considerably

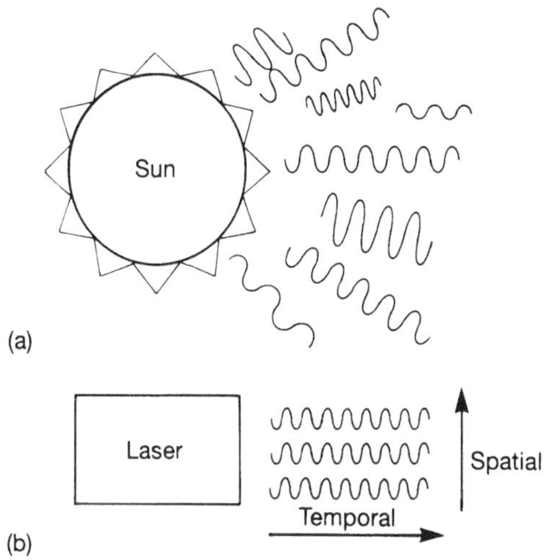


**1.1** Coherent and incoherent light: a) The random light waves from the sun are incoherent because they cannot achieve temporal and spatial symmetry; b) The coherent light waves from a laser.

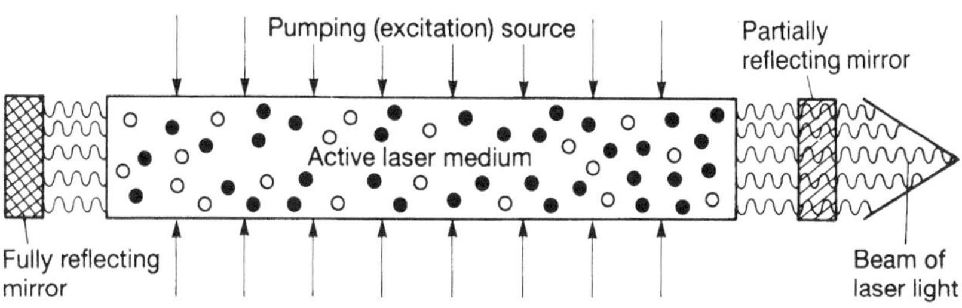


**1.2** The basic elements of a laser.

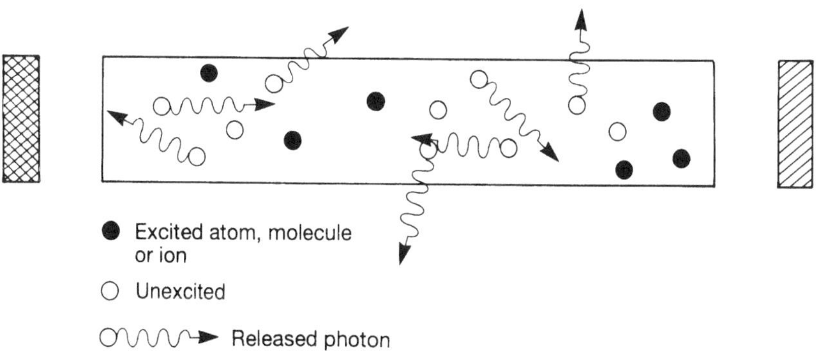


**1.3** Spontaneous emission of photons from the excited active medium. There are billions of excited atoms, molecules and ions and they release their photons in all directions.

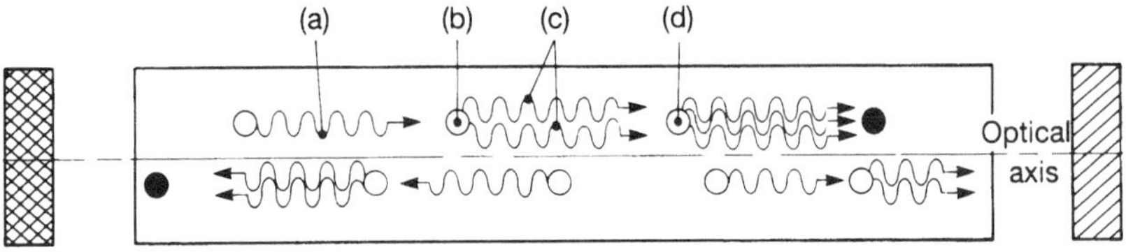


**1.4** Stimulated emission of photons. A photon (a) which collides with an excited atom, etc (b) will cause it to release its photon before spontaneous emission can occur and thus two photons (c) will travel on in phase until the next collision (d). By stimulated emission photons which travel parallel to the optical axis build a powerful laser beam.

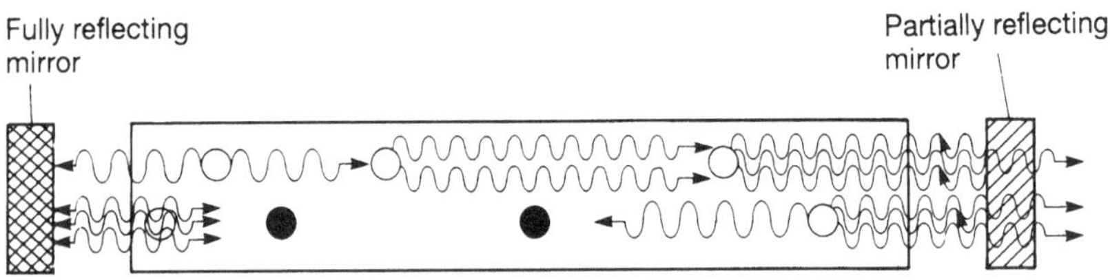


**1.5** Optical feedback of photons by mirrors to increase the path length for stimulated emission and thus amplify the laser power.

extended by the optical feed back provided by the mirrors, before leaving the laser, through the partially transmitting mirror, Fig. 1.5. This action not only serves as an amplifier for photon generation by stimulated emission to achieve the required power level, but also to provide the highly collimated coherent light beam that makes the laser so useful.

The power density across the diameter of a laser output beam is not uniform and is dependent on the laser's active medium, its internal dimensions, optical feed back design and the excitation system employed. The transverse cross sectional profile of a laser beam, which shows its power distribution, is called the 'transverse electromagnetic mode' (TEM). Many different TEMs can be designed for and each type is rated by a number. In general, the higher the number the more difficult it is to focus the laser beam to a fine spot to achieve a high power density. Because the latter is paramount when laser welding, lasers with $TEM_{00}$, $TEM_{01}*$, $TEM_{10}$, $TEM_{11}*$ and $TEM_{20}$ and combinations of these modes are often used. Figure 1.6 shows the basic shape of the beam power profiles of these modes. Some lasers produce several different modes and these are usually referred to as having a multi-mode operation.

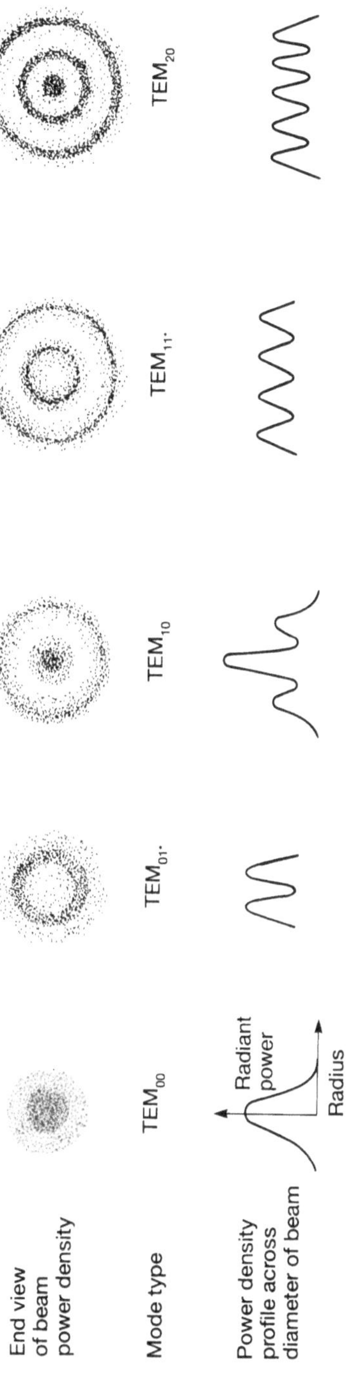


**1.6** Basic beam modes produced by different lasers; some lasers have combinations of these modes.

## Welding lasers

The present welding lasers, which can also be used for cutting and surfacing materials, use active mediums which are either in the form of a solid or a gas. Consequently, the two types are referred to as 'solid-state lasers' and 'gas lasers'.

Three main types of solid-state laser have been developed: the ruby, neodymium glass and the neodymium yttrium aluminium garnet (Nd:YAG). The Nd:YAG, which has an output wave length of 1.06 μm, has practically replaced the other two types where high production welding is required, as it can achieve higher powers for longer periods, without overheating and thus degrading its performance. Therefore, with respect to solid-state lasers this book will concentrate on the Nd:YAG laser.

The gas lasers which are used for welding are currently all 10.6 μm wave length carbon dioxide ($CO_2$) lasers since they have proved to be the most efficient and produce the highest power. However, carbon monoxide (CO) lasers which have a shorter wave length, 5.3 μm, are under development and evaluation.

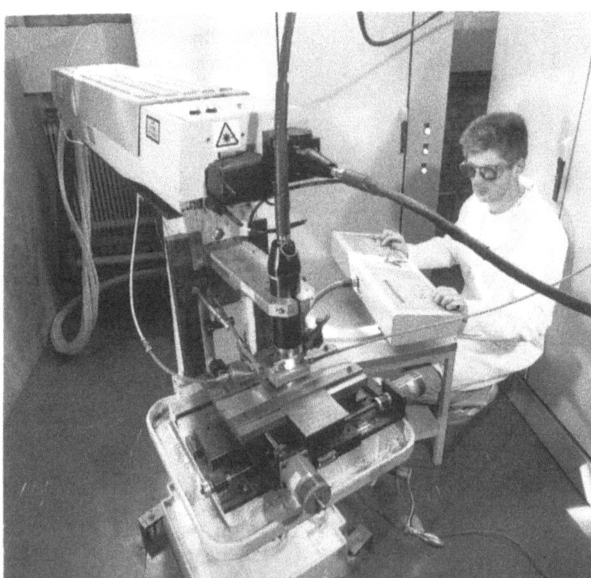

**1.7** A 1 kW Nd:YAG laser. The beam is transmitted from the laser via an optical fibre to the laser gun, centre foreground. Behind the laser and operator are the laser power supply cabinets.

## Nd:YAG lasers

Nd:YAG lasers with output powers (average) ranging from approximately 100 W to over 1 kW are commercially available; higher power machines of 2 to 3 kW are being developed. A 1 kW Nd:YAG welding laser and its power supply is shown in Fig. 1.7. Although the average powers developed by Nd:YAG lasers are low when compared with $CO_2$ lasers, they can achieve pulses with peak powers in the order of 10 kW. Performance-wise a 150 W Nd:YAG laser will butt weld 0.5 mm thick steel at 0.3 m/min and a 1 kW machine will easily butt weld 4 mm at the same speed; typical welding performance figures are presented in Chapter 5.

Perhaps the most important feature of the Nd:YAG laser is that the output wave length (1.06 μm) of its beam can be transmitted through a fibre optic cable: beam transmission systems are discussed in Chapter 11. This aspect makes the Nd:YAG laser extremely attractive for high speed welding production and automation, which are discussed in Chapter 10. However, special safety precautions (Chapter 12) have to be taken due to the short wave length.

A schematic layout of an Nd:YAG laser is shown in Fig. 1.8. The combined assembly shown of flash lamp, laser rod, reflector and mirrors is usually referred to as the optical cavity or resonator. The laser rod, an yttrium aluminium garnet (YAG) crystal, is implanted with carefully distributed neodymium (Nd) atoms. The rod is transparent to the excitation radiation from the flash lamp and also to the photons generated when the Nd atoms are excited. The rod, flash lamps and reflector are immersed in cooling water in a chamber which forms part of the resonator housing. It is necessary to cool the rod continually to avoid thermal distortion, which would degrade the laser performance, and also to prevent the rod from fracturing.

The fully reflecting and partially transmitting mirrors used in the Nd:YAG laser resonator are made of glass and use a dielectric reflective coating.

The power supply (Fig. 1.8.) generates current pulses of the desired amplitude and duration and feeds them into the flash lamps which convert the current pulses into pulses of 'white' light. This light is coupled into the laser rod with the aid of the reflector assembly (the 'pumping chamber') thus exciting the Nd atoms to produce light by spontaneous and stimulated emission. The pulsed excitation of the laser rod results in the generation of a pulse of laser light essentially of the same pulse

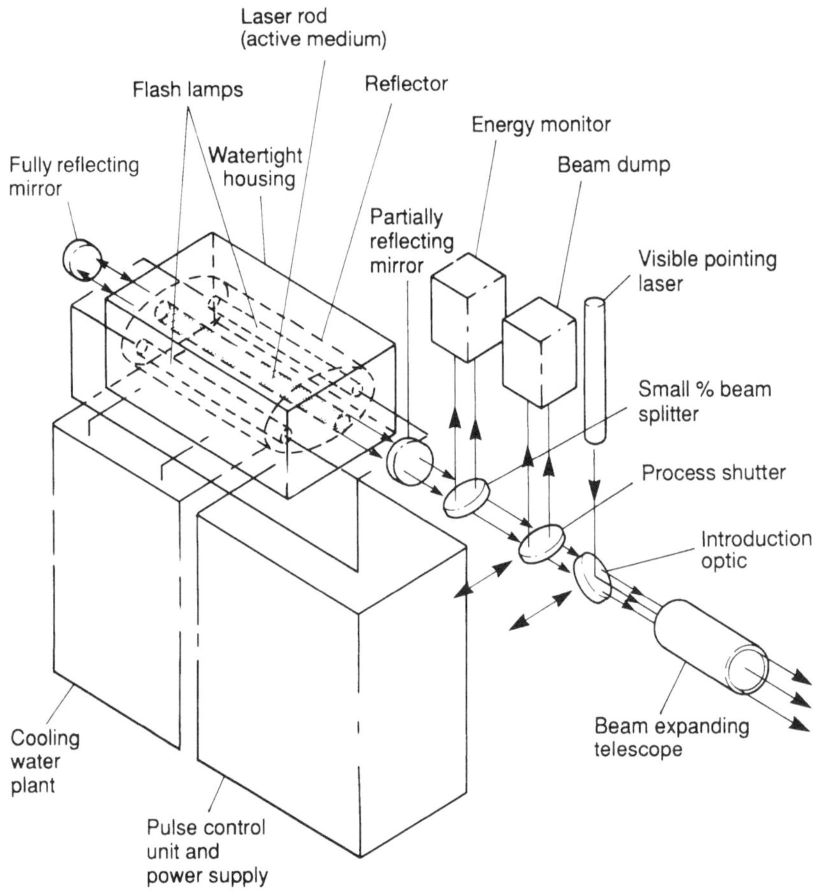


**1.8** Schematic layout of an Nd:YAG laser.

duration as the current pulse from the power supply (not all solid-state lasers are pulsed, continuous wave Nd:YAG lasers have been developed but at present are not widely used for welding).

The ability to control the current pulse parameters allows control of the laser beam/material interaction and hence weld penetration depth, profile and appearance. Typical pulse durations for an Nd:YAG welding laser range from 0.5 to 20 msec and the repetition rate from 5 to 500 Hz.

1.9 A 5 kW axial flow $CO_2$ laser. In the foreground the operator is setting up a component to rotate under the laser beam. The cabinets on the left house the laser power supply.

1.10 A 10 kW cross flow $CO_2$ laser, extreme left, with a beam transmission tube leading into the workstation, right. The laser power supply cabinets are behind the workstation.

## $CO_2$ lasers

$CO_2$ lasers are commercially available with output powers of 0.5 to 25 kW. A 5 kW and a 10 kW laser are shown in Fig. 1.9 and 1.10 respectively. Due to the progress in developing $CO_2$ lasers of increasing power, machines in the 2 kW to 5 kW range are presently used in the largest numbers for welding. A 2 kW laser will achieve a weld depth of 3 mm at 1 m/min in steel whereas a 5 kW and 10 kW laser will comfortably achieve 7 mm and 15 mm respectively for the same speed. More detailed information on $CO_2$ laser welding performance figures is presented in Chapter 5.

A schematic layout of a $CO_2$ gas laser is shown in Fig. 1.11. This type of laser is larger, more complex and can achieve much higher powers than the Nd:YAG lasers described previously. The active $CO_2$

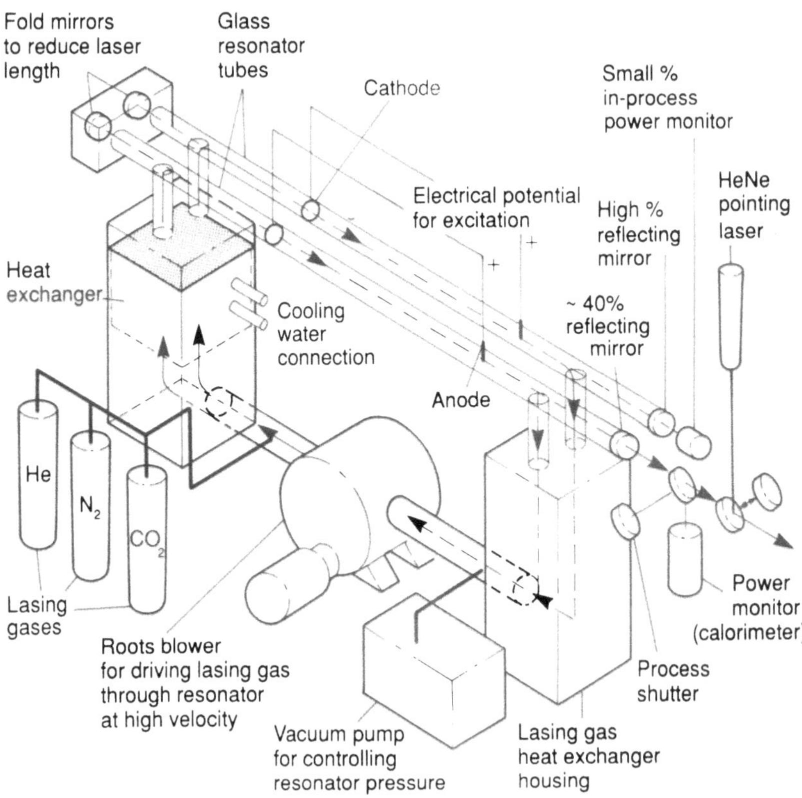


1.11 Schematic layout of an axial flow $CO_2$ laser.

medium, which is assisted in its lasing operation by the addition of helium (He) and nitrogen ($N_2$), is contained at a low pressure of typically 50 mbar in a chamber which forms the resonator housing. The He and $N_2$ when added to the $CO_2$ help remove unwanted wave lengths and promote more efficient photon generation. Prior to filling with the gas mixture, air is removed from the chamber by a vacuum pump which also doubles to control the chamber pressure in relation to the incoming laser gas by slowly bleeding off gas degraded by the continual excitation during the lasing action. The carefully controlled low pressure is necessary to obtain a glow discharge from the electrical potential, which is applied to excite the gas molecules which occupy much of the volume of the optical path between the resonator mirrors.

Both direct current (DC) and radio frequency (RF) high voltage systems can be used to excite the gas lasers and the input power required is typically ten times that of the laser output. The current can be applied either in the continuous wave or pulsed mode. The former is the most common.

The gas within the laser is driven in a continuous loop through the resonator and a heat exchanger system to keep it cool and make the lasing action more efficient. Two different types of gas flow loop are employed in $CO_2$ lasers; they are referred to as axial flow, where the laser gas flows co-axial with the optical axis, as shown in Fig. 1.11, and cross flow, where the flow is at 90° to the axis, Fig. 1.12. Axial flow lasers generally produce lower TEM number beam modes than cross flow lasers; $TEM_{00}$ or $TEM_{01}$*. However, cross flow lasers can achieve higher powers for a given resonator length, which enables more compact lasers to be made at the high power end of the $CO_2$ laser range. In general, axial flow lasers dominate the market for laser powers up to 5 kW and cross flow lasers from 5 to 25 kW.

The resonators of $CO_2$ gas lasers are often referred to as 'stable' and 'unstable' resonators. This unfortunate terminology does not mean that one performs better than the other but rather to the method by which the laser power is extracted from the resonator. A stable resonator is one in which a percentage of the laser light, typically 40%, is taken out uniformally over the area of a partially reflecting window (mirror) and applies to almost all axial flow lasers. In the unstable resonator a similar percentage of light is extracted by a scraper mirror, Fig. 1.13, angled at 45° to the optical axis of the laser resonator and between two fully reflecting mirrors. The scraper mirror has a hole in it which is co-axial with the resonator beam and therefore only scrapes

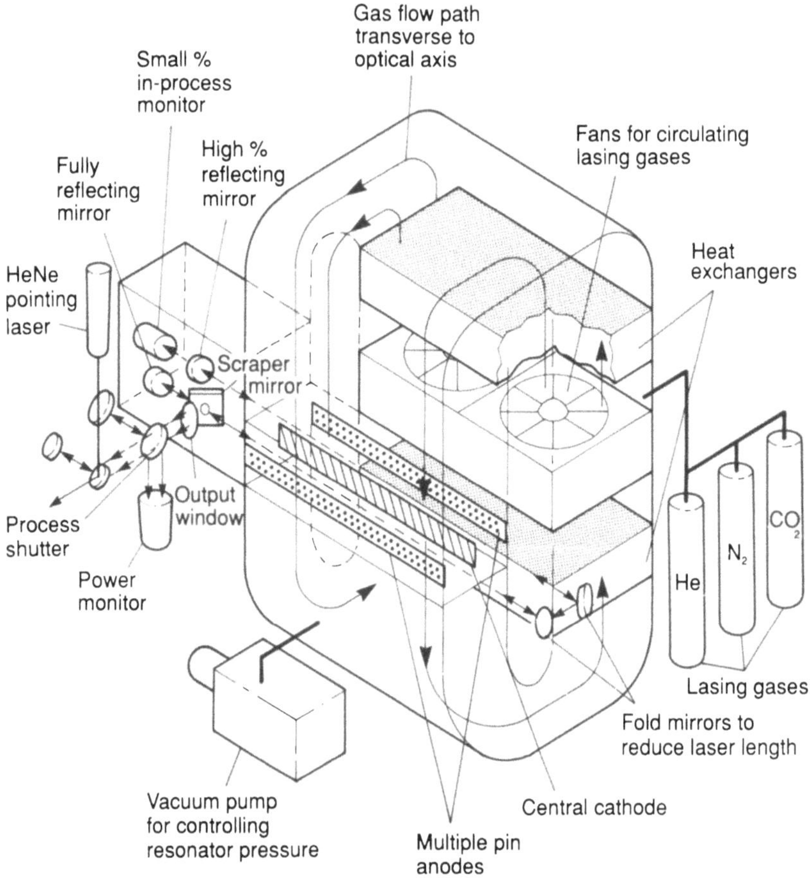


**1.12** Schematic layout of a cross flow $CO_2$ laser.

off the outer annulus of the resonator beam. The fully reflecting mirrors are configured so that as the beam is reflected backwards and forwards in the resonator it gradually propagates and spills over on to the scraper mirror. Scraper mirrors are generally used on lasers which deliver powers in excess of 5 kW and are usually used in conjunction with an aero-window (Chapter 11). This is because the present partially reflecting mirrors cannot transmit power levels much above 5 kW without thermally distorting and upsetting the transmission of the laser beam. Consequently since most lasers above the 5 kW level are cross flow lasers, these lasers fall into the unstable resonator category.

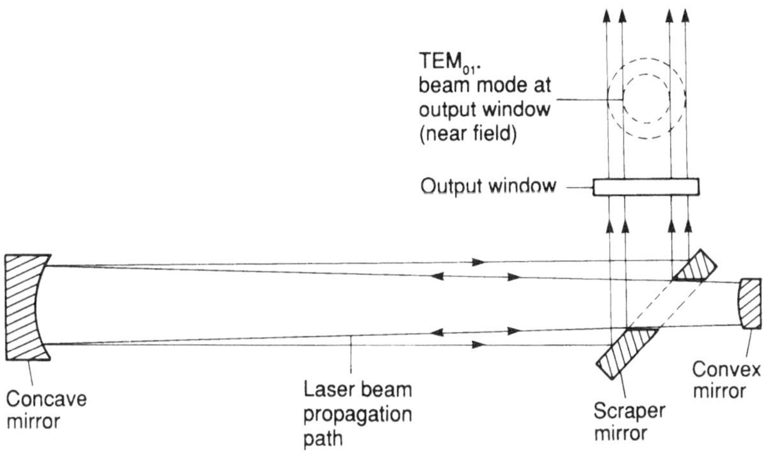


**1.13** Principle of the scraper mirror used, in an unstable resonator, for extracting power from high power (usually >5 kW) $CO_2$ lasers.

## Harnessing the laser beam for welding

Laser output beam diameters increase with the power rating of the laser, for example 1, 5, 10 and 25 kW lasers have beam diameters in the order of 10, 25, 40 and 70 mm respectively. The average power density across these diameters is in the order of 6 to 13 W/mm$^2$; the actual concentrations of power being distributed according to the beam mode (see Fig. 1.6). For laser 'keyhole' welding (Chapter 2) power densities in the order of $10^3$ to $10^5$ W/mm$^2$ are required and therefore it is necessary to focus the laser output beam to a very small spot of only a fraction of a millimetre in diameter.

The almost parallel beam from a laser cannot be focused to infinity before it diverges again. The focused spot size is determined by the laser beam diameter, the focal length of the focusing optic used, the beam mode and the beam divergence angle (the divergence angle is the angle at which the almost parallel laser beam propagates on leaving the laser).

There are numerous methods of calculating the focus spot size of a laser beam in order to ascertain the potential power density of a laser. Physical measurement is not easy, especially at high laser powers, since it requires scientific equipment which is not always practical on the shop floor. However, a satisfactory guide to the approximate focused spot diameter for most types of $CO_2$ laser (especially those with stable resonators) can be obtained from the formula:

$$\text{Focused spot diameter} = \frac{2 \cdot 44\lambda F}{D} \quad (2M + 1)$$

Where $\lambda$ = laser light wave length (mm).
$F$ = the focal length (mm) of the focusing optic used.
$D$ = laser beam diameter (mm) at the focus optic.
$M$ = laser beam mode number expressed as a decimal, with the decimal point placed before the first digit e.g. $TEM_{01} = 0 \cdot 01$ and $TEM_{20} = 0 \cdot 2$.

*Example*
A $CO_2$ laser (10.6 µm wave length) with a $TEM_{20}$ beam mode and a beam diameter of 38 mm at the focusing optic, when used in conjunction with a 300 mm focal length optic will produce a focused spot size of:

$$\frac{2.44 \times 0.0106 \times 300}{38} \times (2 \times 0.2 + 1)$$

$$= \frac{7.7592}{38} \times 1.4$$

= approximately 0.29 mm diameter.

Nd:YAG welding lasers generally have larger beam divergence angles than $CO_2$ lasers and therefore cannot be focused to such small spot sizes without using a collimator placed before the focusing optic (a telescope in reverse). The approximate focused spot size for these lasers is usually estimated from the formula:

Focused spot diameter = $2\,\theta\,F$
where $\theta$ = the divergence angle (radians) of the laser beam as it leaves the laser or collimator.
$F$ = the focal length (mm) of the focusing lens used.

In Chapter 5, it will be shown that establishing the focus spot size by calculation is not as important as it may at first appear. From a practical standpoint the focusing f number is more useful to establish tolerant welding conditions. The f number is derived by dividing the focal length of the focusing optic by the laser beam diameter applied. The selection of focus f number/spot size is discussed in Chapter 5.

Transmissive optics (lenses) and reflective optics (shaped mirrors) are used to focus laser beams. The different types and the selection of optics is discussed in Chapter 11.

## References

1 Koechner W: 'Solid state laser engineering', 2nd Edition, Springer Verlag, Berlin, Heidelberg and New York.
2 Duley W W: '$CO_2$ lasers - Effects and applications', Academic Press, London and New York. ISBN 0-12-223350-6.

Chapter 2

# Principles of laser welding and its characteristics

This chapter describes the formation of a laser weld; the characteristics which distinguish laser welding from other fusion welding techniques; and the necessary considerations when choosing laser welding.

## Laser welding

Laser welding is in general a keyhole fusion welding technique which is achieved with the very high power density obtained by focusing a beam of laser light to a very fine spot. As with electron beam (EB) welding, it is also often referred to as a high energy density or power beam technique.

To form a laser weld, the laser beam is brought to focus on or very near the surface of the workpieces to be joined. In the first instance a large percentage of the incident beam is reflected from the work surface for a minute period; this is because most metals are good reflectors. However, the small amount of laser beam energy which is initially absorbed by the work quickly heats the material surface, causing production of an energy absorbing ionised metal vapour, which rapidly accelerates the absorption of much of the energy that previously would have been reflected.

At a focused power density in the order of $10^4$ W/mm$^2$ the rapid removal of metal by vaporisation initiates a small keyhole (a very small diameter cylindrical shaft) into the workpiece. As the keyhole penetrates deeper into the workpiece the laser light is scattered repeatedly within it, thus increasing the coupling of laser energy into the work. While the laser energy is applied, the keyhole, Fig. 2.1, is held open by vapour pressure which prevents the molten walls from collapsing. If the laser is operated in a pulsed energy mode the molten metal will slump to the centre of the keyhole and solidify on completion of the pulse, sometimes leaving a slightly undercut weld bead due to the loss

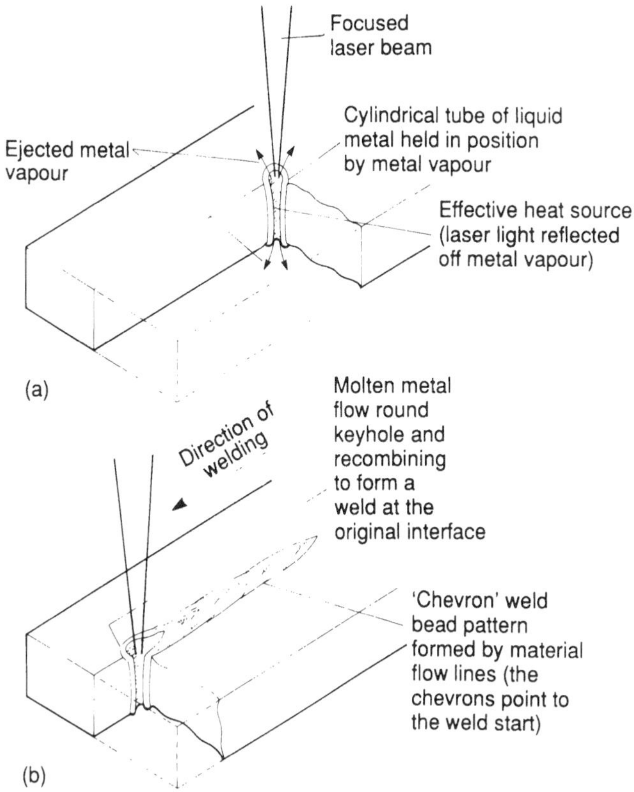


**2.1** a) A laser weld keyhole; b) Its movement through the workpiece to recombine molten metal at its trailing edge to form a continuous weld.

of metal through vaporisation when the keyhole was first formed. When the laser is operated in the continuous wave mode i.e. not pulsed, and the beam is moved along the joint line, or vice versa, the molten material at the leading edge of the keyhole produces a very small wave which is supported by surface tension. As the beam travels forward the wave reorientates round the edge and slumps to the back of the keyhole as the molten walls at the trailing edge come together and solidify through the depth of the keyhole. This action leaves, on close fitting joints, a slightly protruding top surface weld bead with a chevron pattern which points towards the start of the weld. A proportion of the weld bead height is also produced by axial shrinkage of the weld metal in the direction transverse to the direction of welding. The top weld beads are generally neat and have an aesthetic appearance in comparison to conventional arc welds.

Laser keyholing will initially occur in steel at a power density of about $10^3$ W/mm$^2$. At this level the welds tend to be shallow and wide, Fig. 2.2, like plasma arc keyhole welds and require long beam to work-piece interaction times for their formation. In contrast, at power densities in the order of $10^4$–$10^5$ W/mm$^2$ the welds are deep and narrow requiring short interaction times, thus promoting high welding speeds (see Fig. 2.3, 2.4 and 2.5). At power densities marginally greater than $10^5$ W/mm$^2$ cutting and drilling regimes are reached and weld metal loss occurs through expulsion of molten and evaporated metal, producing weld undercutting and eliminating a protruding weld bead on the top surface of the work. Nevertheless, excessive power rarely produces blowholes in the weld, especially when welding thin sections at high speed. This is because much of the excess power disappears through the bottom of the keyhole.

## Characteristics and considerations

As laser welding is a high energy density welding process it does not depend solely on heat conduction for achieving weld penetration depth and is therefore quite different to conventional conduction limited gas and arc welding. With these two techniques weld fusion and penetration depth is achieved as heat is conducted into the work, Fig. 2.3, and the melting point isotherm (a line connecting points at an equal temperature) moves outwards in all directions from the source. The widths of welds produced in this manner are usually greater than their depth and the heat input is more than is strictly necessary to

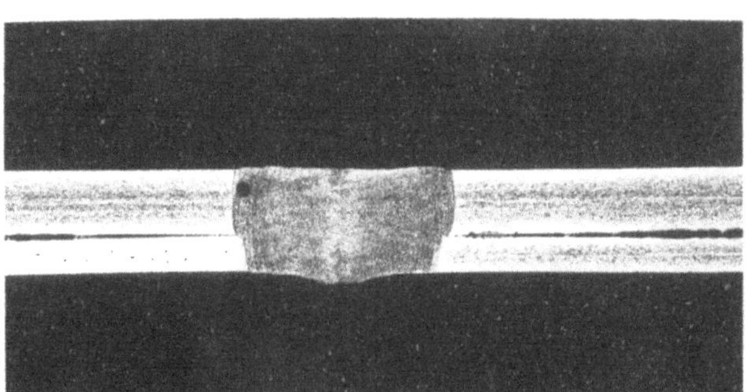

**2.2** A transverse cross section through a laser welded lap seam joint made at a power density in the order of $10^3$ W/mm$^2$, which is bordering on the keyhole welding threshold. The materials are type 304 stainless steel, 1 and 0.5 mm thick.

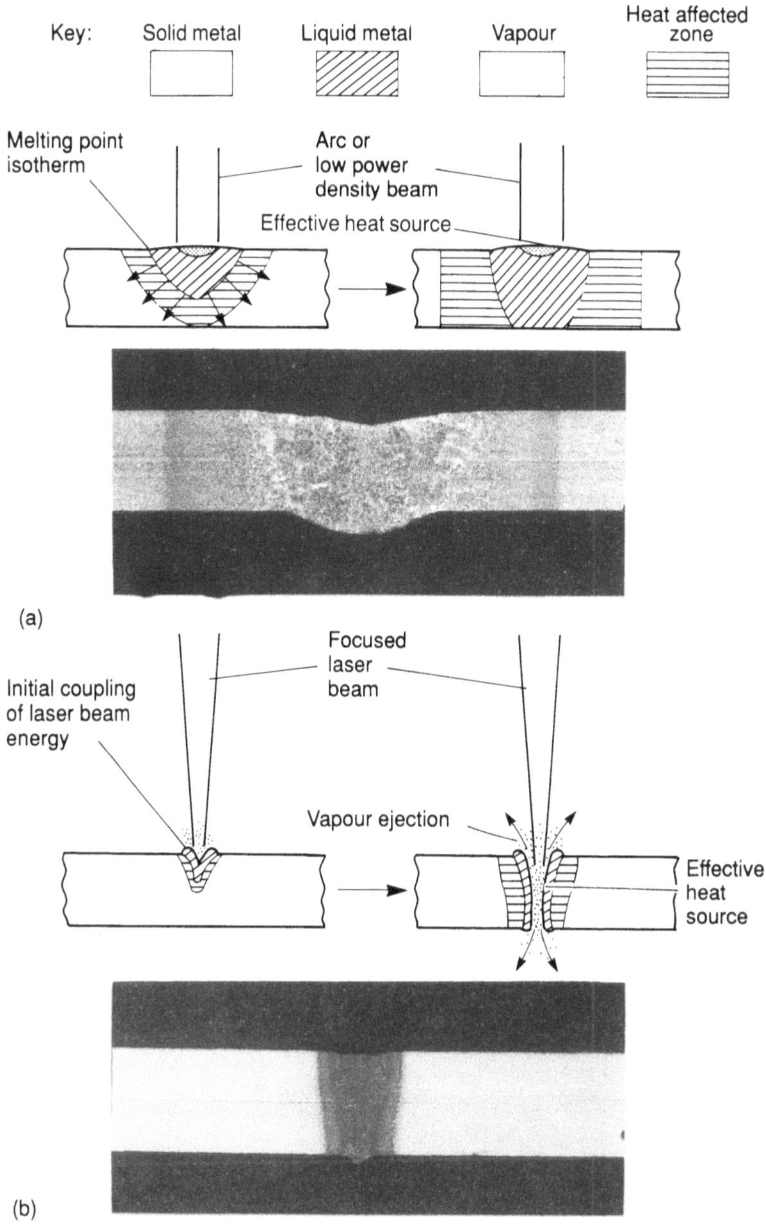


**2.3** The difference between (a) a conduction limited weld formation and (b) a keyhole formation. The transverse butt weld cross section of a conduction limited weld was made by tungsten inert gas welding. This weld and the laser weld cross section shown are both made in 3 mm thick mild steel.

achieve the required weld depth. Above a few millimetres the penetration depth of gas and arc welding is limited and the joint has to be given a V groove preparation. This has to be filled with melted filler wire or rod and by a number of separate overlapping weld passes.

The laser keyhole welding technique transfers heat from the laser source into the material not to just a point on the surface, but to a line extending through the material thickness, Fig. 2.4. The weld penetration depth in a close fitting butt joint is therefore, in general, only limited by the laser power available. By rule of thumb 1 kW of correctly focused laser light will comfortably weld to a depth of 1.5 mm at a speed of 1 m/min when welding steel; thereby 10 kW will weld 15 mm. Consequently, much thicker sections can be welded in a single pass than by gas and arc welding thus reducing production times and eliminating the cost of filler wire. This is illustrated in Fig. 2.5, where a single pass laser weld from each side has been superimposed over the multipass arc welded butt joint.

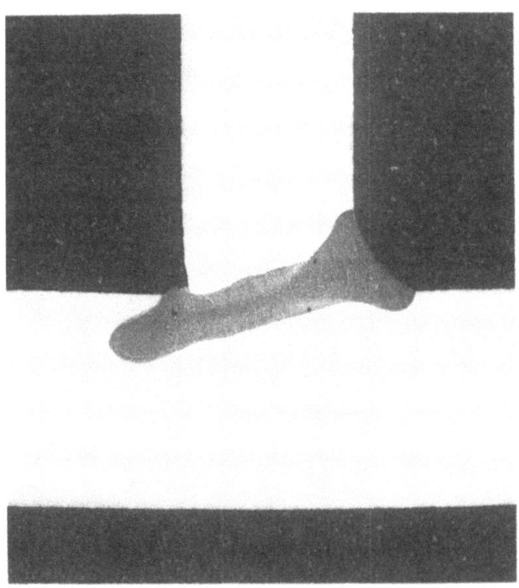

**2.4** A section through a laser keyhole weld showing the heat transfer into the material and not to just a point on the work surface as in a conduction limited weld. The T butt joint weld shown is in 3.2 mm thick C263 nickel alloy and was made at a laser power of 5 kW, a welding speed of 3 m/min and a laser beam impingement angle of 15° to the horizontal plate.

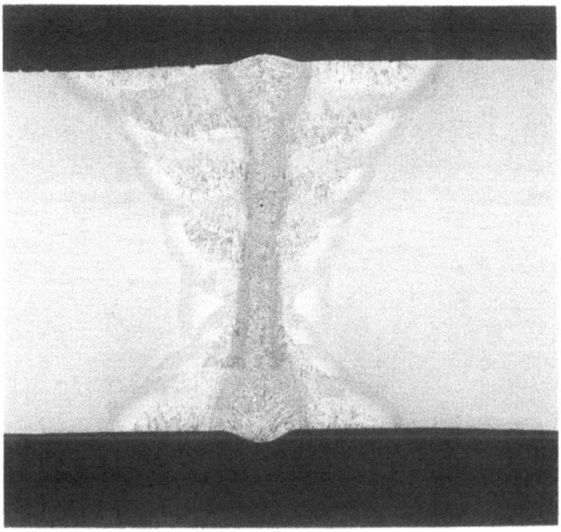

**2.5** A single pass laser weld from each side (two passes) of a 25 mm thick steel plate superimposed on a multipass (11 passes) manual metal arc weld.

## Advantages

Laser welding with a beam power density of $10^4$ W/mm$^2$ at the work surface is extremely advantageous, often permitting welding speeds of several metres per minute and a total heat input which is much lower than that achieved when arc welding a similar workpiece. Added to this, the deep narrow and near parallel laser weld shape, plus the very narrow heat affected zone (HAZ) collectively produce very little workpiece distortion.

Because the laser weld keyhole can extend as a deep narrow line through a material, it provides other advantages besides high welding speeds and low workpiece distortion. The deep penetration capability allows several layers of material to be lap welded in a single pass. A section through a weld joining four layers of 0.7 mm thick low carbon steel is shown in Fig. 2.6. Another advantage is that the weld can usually be placed exactly where it is required, i.e. at the point of workpiece abutment. This aspect is particularly important, because a weld which joins parts through the full thickness of the abutting faces is strong in both fatigue and tensile strength, as it allows the lines of stress to pass through the joint smoothly. In a joint made with a partial penetration

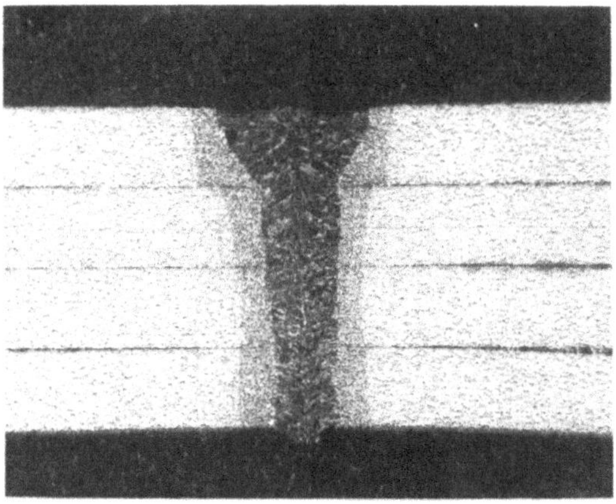

**2.6** A transverse cross section through a four layer lap seam weld between steel sheets, each 0.7 mm thick, illustrating another example and advantage of the deep penetration capability of laser welding.

butt weld, or with a fillet weld, the lines of stress are bent in their passage through the joint, producing stress concentrations which reduce fatigue strength. Typical cases are the corner, Fig. 2.7, and T butt joint which are normally joined by a fillet of added filler wire when using arc welding.

The narrow tapering laser beam as it comes to focus offers design flexibility. It can often access joint configurations that are inaccessible to other welding techniques, such as butt joints between gears, Fig. 2.8. The ability to access narrow gaps eliminates some important joint design considerations such as land sizes for resistance welding electrode connections, potential arc short circuit paths when arc welding and magnetic fields when EB welding. The accessibility offered by the laser beam also enables components to be fabricated where access is only available or practical from one side, such as in fabrication of sandwich panels containing a corrugated core. This application is illustrated in Chapter 6.

Further process flexibility is achieved with the unfocused laser beam which on leaving the laser can be transmitted several metres without serious loss of power. Therefore, several workstations can time share one laser, Fig. 2.9, enabling higher production of one component or a range of different components. Furthermore, the laser beam can be manipulated about an irregular shaped workpiece by using a robot. Multiple workstations and beam transmission systems are described in Chapter 10 and 11 respectively.

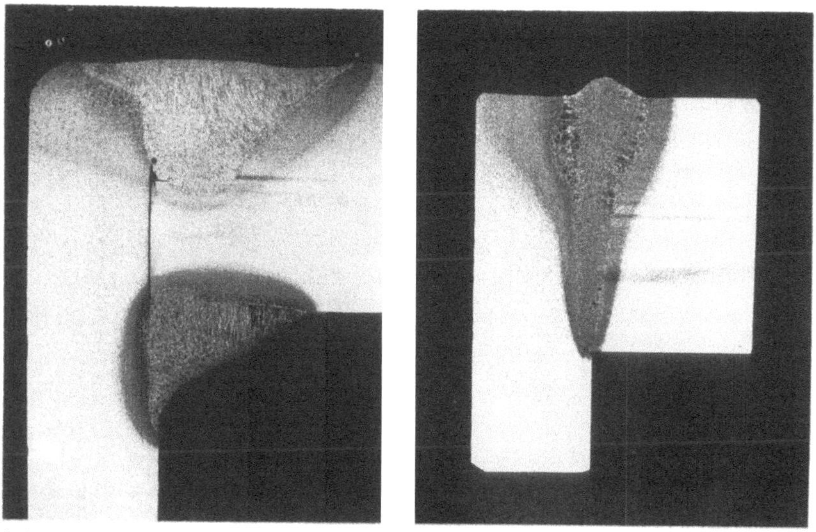

(a) (b)

**2.7** Examples of 18 mm deep edge butt joints in carbon-manganese steel, made by (a) arc welding and (b) laser welding.

**2.8** Circumferential $CO_2$ laser welds being made between the abutting faces of gears, where access to the joint is limited.

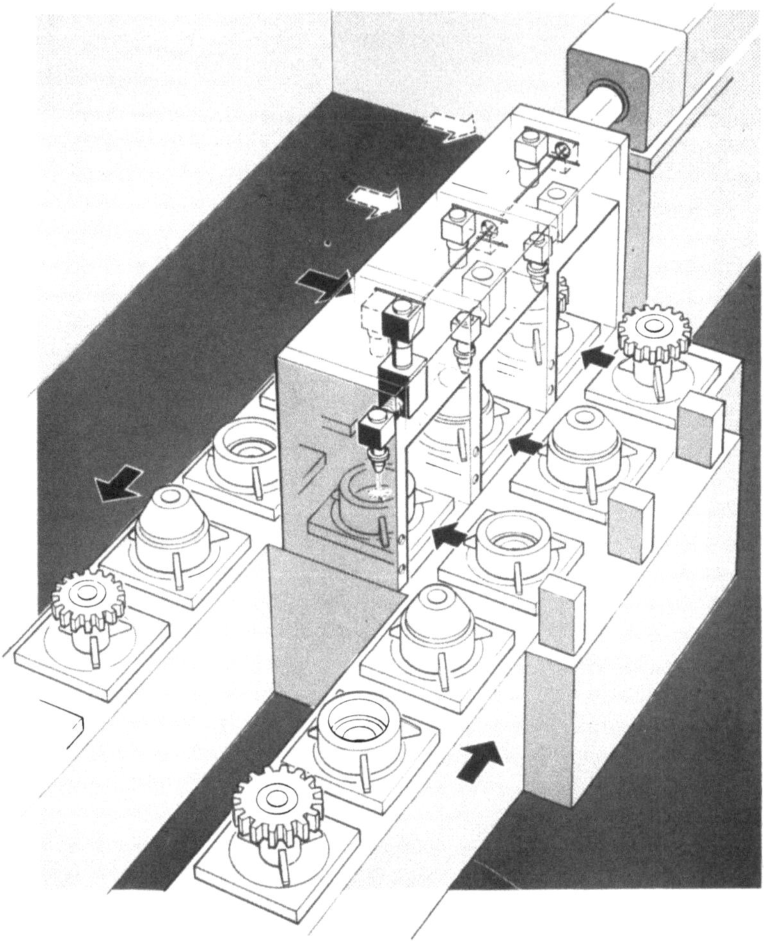

2.9 Schematic arrangement of three workstations time sharing one laser.

## Welding positions

For operator safety, and that of other personnel, the laser gun has to be mechanically manipulated if it is required to move about a stationary workpiece. Nevertheless, it can be operated in all welding positions from the down hand to the vertical above. However, unless a sacrificial window is used at the exit of the laser gun, it is normal to operate the gun between the down hand and horizontal position to avoid weld spatter and other debris falling back on to, and damaging, the focusing optic. The ability to move into the overhead position is due to the high surface tension in the very narrow weld keyhole and the welding speed (high rate of cooling) preventing the molten metal from dropping out under the force of gravity. For the same reason, weld backing bars are not necessary when laser welding is in the vertical down position.

## Materials

Lasers can weld almost the same range of materials as other fusion welding techniques, however, aluminium and copper and their alloys do present difficulties to laser welding. This is mainly because these materials do not easily absorb laser light because they are good reflectors. In fact, copper reflects 10 µm light wave lengths so well it is used for $CO_2$ laser mirrors. As a general guide, the lower the thermal conductivity or the higher the electrical resistivity of a material the more likely it is to absorb laser light. Consequently, the normal weldable grades of steel and stainless steel are ideal for laser welding. A basic guide to materials which can be joined by laser welding is given in Chapter 4.

## Joint configurations

Laser welding can be applied to most of the basic joint configurations as shown in Fig. 2.10. There are other more complex shapes that can also be laser welded; Houldcroft in his book 'Which Process?'[1] illustrates 26 suitable joints which are recognisably different. In this book, Chapter 6 highlights joint configurations which are generally used for joining sheet metal parts such as those used in automotive body and domestic appliance construction. Chapter 7 covers joints used in plate and tube assemblies as used in heavier engineering applications. Perhaps, of the common engineering joint configurations, the two most difficult to laser weld are the butt joint between two solid round bars and the T in round pipe joint. In the butt joint between solid bars the welding velocity reduces as the weld keyhole nears the centre of the bar. This produces excessive heating at the centre of the joint and the risk of serious material expulsion and resultant weld porosity. Solid round bar butt joints are therefore better welded by resistance butt, flash butt or friction welding techniques; these are described in Ref. 1. The T in round pipe joint presents part fit-up and joint line following difficulties. In large fixed structures the latter is impossible to achieve without a robot. Both difficulties are, however, easily overcome using manual arc welding techniques and the addition of a filler material. The example of the T in round pipe joint also serves to make the point, that the joint type and component size play an important part when choosing laser welding. Therefore, one of the early considerations is to establish if it is possible, practical and cost effective to manipulate the work about the laser beam or vice versa.

### Disadvantages

So far, characteristics which in the main have positive attributes have been described. However, as have all welding techniques, laser welding has its disadvantages. The focused spot size of a laser beam, which is only a fraction of a millimetre in diameter, requires close fitting joints (unless the joint is a lap joint or filler wire is used) otherwise a large proportion of laser beam energy will be lost through any gap between the joint faces. Moreover, because the fusion zone width produced in a laser weld is so narrow, great care has to be taken in aligning the laser beam with the joint line. (Joint fit-up and beam/joint alignment tolerances together with advice on how to establish them is given in Chapter 6 for thin sheet and Chapter 7 for plate.)

Good dimensional quality workpieces and workpiece laser beam manipulation equipment is paramount for laser welding. This is necessary not only to achieve good beam/joint line alignment, but also to control the beam focus position and welding speed and hence the energy input into the work. Annular and circumferential butt joints which can be rotated concentrically with the axis of a rotary table can usually be welded with simple and inexpensive equipment. On the other hand, long straight butt joints in large thin sheets require expensive high precision alignment and clamping systems. Likewise, components with irregular joint lines such as the T in round pipe joint, require expensive precision made, computer controlled multi-axis equipment.

Welding lasers, especially the $CO_2$ gas lasers are not portable in the sense that they can be taken on site. Small lasers have been mounted on robot arms and gantry systems, which makes them semi-portable, but they are still workshop or production line based machines. There are perhaps three main reasons for this: 1) Lasers ideally need to be mounted on a stable base to maintain the optical alignment of their resonators; 2) To operate they require large electrical power supplies and relatively large cooling water plants (Chapter 13); 3) At all times special safety screening (Chapter 12) encompassing the working envelope of the laser gun, must be provided.

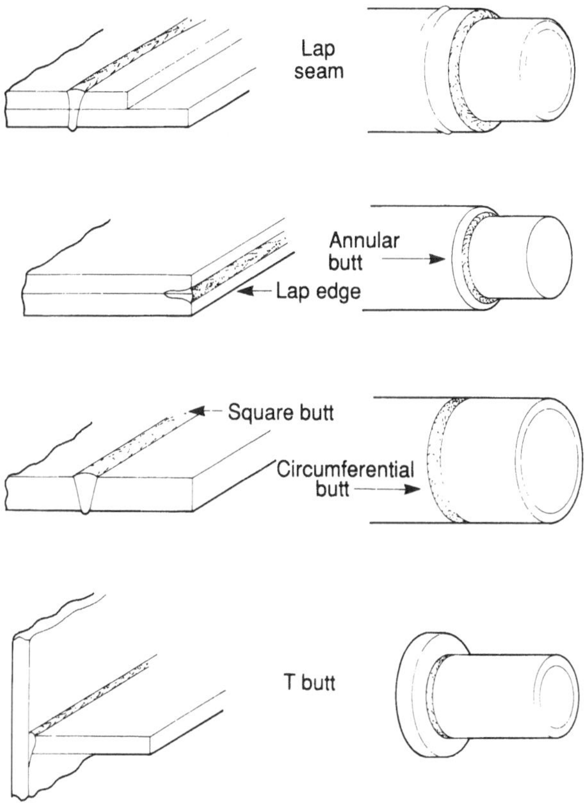


**2.10** The basic joint configurations which can be laser welded.

In comparison to arc and many other types of welding machines, lasers with the necessary work handling or beam manipulation equipment are expensive. Houldcroft[1] estimates that a general purpose laser can cost from 50 to 200 times more than a conventional industrial arc welding machine and a high power special purpose laser system as much as 100–750 times more. The low estimate would cover a small low power laser with a simple workpiece manipulator, such as a rotary table. Laser operating costs are also high and are mainly due to their poor power conversion efficiency (typically 5–10%) in relation to the input power required to drive them. The high capital and operation costs are in most cases offset by the lasers very high weld productivity. Nevertheless, to keep production costs down high laser utilization in terms of welding throughput is essential.

At this point it should be noted that laser welding can often eliminate up and down stream machining operations and enable a reduction in component parts to be achieved. These aspects can lead to significant cost savings which can be set against the laser operating costs. Examples of reductions in machining and component parts can be found in Chapter 3.

From the general characteristics of laser welding which have been described so far, and for quick reference at a later date, the major advantages and disadvantages of laser welding are drawn together and presented in Table 2.1. The main features are presented as headings and important aspects which are pertinent to each feature are presented beneath.

***Table 2.1*** **Advantages and disadvantages of laser welding**

ADVANTAGES

1 Deep narrow welds
- Single pass weld penetration depth is only limited by the laser power available. (A 10 kW laser will weld 15 mm thick steel in one pass.) This can eliminate the need for a V joint preparation and the addition of a filler material.

2 Low heat input in material each side of the welded joint
- Very low thermal distortion of the workpiece occurs, enabling machined parts to be welded without the need for further post-weld machining.
- Welds can be made very close to heat sensitive components, such as electronic circuits and glass-to-metal seals.
- Metallurgical damage such as severe grain growth and extended HAZ are reduced.

3 High production rates and process flexibility
- Welding speeds are high and can be up to several metres per minute.
- Several workstations can time share one laser.
- Laser welding machines are suited to automation and also operation in conjunction with robots.
- Welding can take place in all laser gun positions, provided a focusing optic protection device is used when working between the horizontal and overhead positions.

4 Welds can be made where access is limited and from one side only
- Welds can usually be positioned exactly at the abutting joint interfaces, thus improving fatigue and tensile strength when compared with alternative arc fillet welds.
- Multiple layers of material can be lap welded from one side and in one pass.
- Welds can often be made at the base of narrow gaps which are inaccessible to other welding techniques.

5 Component design opportunities are enhanced
- A wide range of different joint configurations and dissimilar material thicknesses can be welded, extending design opportunities.
- Material allowances for post-weld machining are often unnecessary and weld land sizes can be reduced, promoting material savings.
- Design considerations concerning magnetic fields and short circuit paths are eliminated.

***Table 2.1 cont'd***

DISADVANTAGES

1 Close fitting and well clamped joints are required
- The small focused spot size of a laser beam will pass through narrow gaps, especially between thin sheets.
- Poorly fitting parts produce undercut welds, unless filler wire is used.

2 Accurate beam/joint alignment is necessary
- The narrow weld can easily miss the joint line if not accurately positioned.
- The depth of beam focus is small and its position about the work surface has to be accurately maintained to achieve the required power density.

3 Precision workpiece or beam manipulation equipment is necessary to control energy input
- The performance of manipulation equipment is also directly related to 2 above.

4 Machines are workshop-based
- Safety screening around the operating envelope of the laser gun is essential for operator safety. This aspect is difficult to achieve for general on-site welding operations.
- The optical stability of the laser resonator and system which transmits the laser beam to the work is paramount for reliable welding performance. Therefore, these equipments need to be maintained on a stable base.
- The electrical and cooled water services required to operate a laser, especially a high power $CO_2$ laser, are not readily portable.

5 Total equipment and operating costs are high
- In comparison with arc welding machines, lasers together with the necessary work handling and ancillary equipment are expensive to purchase and operate. They require high utilization to be cost effective.
- Careful adherence to maintenance schedules is necessary to ensure high equipment up-times.

## Choosing laser welding

Whatever welding technique is used to fabricate a product, it should be carefully chosen at the design stage to obtain the maximum benefit. This unfortunately is easier said than done, because many design engineers are not trained in welding technology and are therefore unaware of the many different welding techniques available (Houldcroft,[1] for example, describes twenty five different welding techniques). Moreover, each technique has attributes which make it particularly suitable for specific welding applications. Nevertheless, from this chapter it is hoped that the reader has built up an understanding of the characteristics, advantages and disadvantages of laser welding and by the end of the book will have a good general knowledge on this subject. With this in mind, design engineers are directed to Table 2.2 where a list of important considerations is drawn up to help establish, at the design stage, if laser welding would be beneficial to the fabrication of a potential product. Depending on the purpose of the product, it is recommended that the list and product be discussed, where necessary, with metallurgical, stress, welding and production engineers. The list, with little modification, can also be used to judge and compare other welding techniques. The significance of Table 2.2 will be realised in the next chapter where a number of different industrial laser welding applications are described and discussed.

***Table 2.2*** **Choosing laser welding – considerations at the product design stage**

Set out below is a question checklist for the designer who has a basic understanding of laser welding and is considering its use for the fabrication of a potential product.

Presented in parentheses are the chapter and reference numbers where, if necessary, information can be found to help answer the questions.

If the answers to questions 1 – 4 are positive, then the remaining questions should be examined to clarify if the equipment and production requirements are practical and also to establish any potential hidden benefits which could improve the product and/or reduce the overall production costs. If any of the answers to questions 1 – 4 are negative, or the answers to questions 5 – 7 are unsatisfactory, then an alternative joining technique should be considered.

*Considerations*

1 Can the materials be reliably welded with an acceptable tolerance to welding parameter variations (Chapter 4)?
2 Will the weld adequately fulfil its service requirements? (If in doubt professional advice from metallurgical and the pertinent mechanical engineers should be sought.)
3 Is the joint configuration and size practical for laser welding? Can the part be manipulated about the laser beam or vice versa (Chapter 10) whilst achieving the required laser beam/joint line alignment tolerances (Chapter 6 and 7)?
4 Can the required joint fit-up tolerances be achieved (Chapter 6 and 7)? If not, would the addition of filler wire be practical (Chapter 8)?
5 Are the potential production numbers, or weld lengths per product run, high enough to justify equipment and operation costs? If not, is the product suitable for sending to a laser job shop, or are there other products which could benefit from laser welding and could time share the same equipment?
6 If very high numbers of welds are envisaged from the outset, will multiple workstations time sharing one laser be more practical than individual lasers with single stations (Chapter 10)?
7 Will laser welding enable a more simple or practical design to be adopted (Chapter 3) leading to:
   a) Material savings;
   b) The elimination or simplification of product upstream and/or downstream engineering requirements;
   c) Improved product quality from both service and visual standpoints?
8 Have alternative welding or joining techniques[1] been considered and has a similar appraisal for these been made?

## References

1 Houldcroft P T: 'Which Process', Woodhead Publishing Ltd. ISBN 1 85573 008 1.

Chapter 3

# Industrial applications

Laser welding is perhaps one of the newest techniques to be adopted by numerous industries for component fabrication. This does not mean that laser welding supersedes previous welding techniques, but that in some cases it offers component fabrication opportunities previously unavailable. The opportunities can embrace a number of different aspects, such as very low weld distortion; aesthetic appearance; simplified design and reduced component size; reduced material wastage; reduced post-weld machining; ease of automation; high production rates; and high production weld numbers in relation to time between equipment maintenance periods. In this chapter examples of 16 industrial applications have been selected which take advantage of one or more of the above opportunities. To emphasise the industrial scope the examples have been drawn from the following industries - electronic, medical, domestic appliance, automotive, specialist tube, machine tool and control valve. Other examples can be found elsewhere in this book.

## Electron guns for television tubes

Figure 3.1 shows a laser welded colour TV gun and Fig. 3.2 the stainless steel components which make up the gun, prior to their assembly. In this example, approximately 130 laser spot welds are made by an Nd:YAG laser. The once traditional electrical component fabrication techniques such as crimping, riveting, soldering and micro-resistance spot welding cannot meet the major fabrication, production and in-service requirements of this component, which are: very low part distortion during fabrication and when working at the operating temperature; ease of joining operation; automated assembly; high production and low process maintenance.

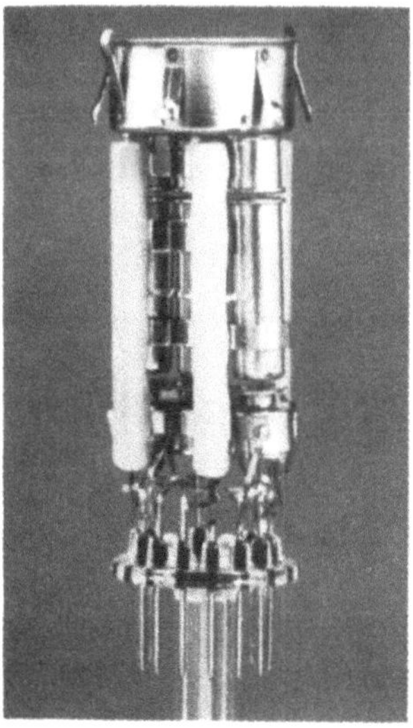

**3.1** A colour television gun fabricated by approximately 130 spot welds made with an Nd:YAG laser (courtesy of Philips).

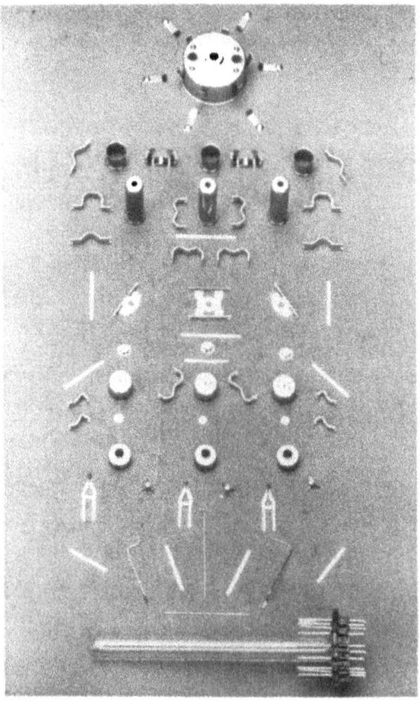

**3.2** The component parts of the colour television gun shown in Fig. 3.1.

The gun assembly is too complex to fabricate by crimping or riveting and to employ these techniques the overall gun size would have to be substantially increased. Furthermore, such joints would be prone to movement at the gun's operating temperature and thus lead to gun failure. There are too many gun parts for hard soldering to be considered. Also soldering would be too expensive, slow and part location extremely difficult. Micro-resistance spot welding the required material thicknesses, which are up to 0.5 mm, would require high electrode forces, which would cause rapid electrode wear and the need for their maintenance. Also, the high stresses produced in the rather large weld heat affected zones could be released as the TV gun assembly heats up during operation, causing distortion leading to poor performance or even failure. Resistance spot welding to the various curved surfaces would offer a high risk of weld splashing and the resultant metal particles could lodge in one of the sensitive parts of the gun assembly and cause complete failure. To have sufficient mechanical strength the resistance spot welding electrodes would need to be large in relation to the overall gun assembly, thus making their access to some joints difficult.

In comparison to the techniques just described, Nd:YAG laser welding offers the following advantages: it is easy to obtain almost splash-free, very low stressed, high quality spot welds between the stainless steel parts; the narrow laser beam can access spot weld sites placed in confined spaces; the weld sites can be small and thus reduce material and component part size; welds can be made at high speed and simultaneously if necessary (Chapter 10); the technique is ideally suited to automation. The drawbacks are the need for very accurate pre-weld part assembly and laser beam positioning and the high overall equipment cost. Despite these drawbacks, in the first five years of laser welded TV gun production[1] laser welding led to reject rates of less than 0.01%, high product quality and when compared with the previous resistance spot welded TV guns, the production maintenance requirement was 70% less.

$CO_2$ laser welding could also be applied to the above application, but it would not be as flexible as Nd:YAG welding for laser beam switching and manipulation reasons, as explained in Chapter 10.

## Razor blades

The Gillette Sensor razor, Fig. 3.3, with its floating blade cartridge, is an excellent example of a laser spot welding application which can

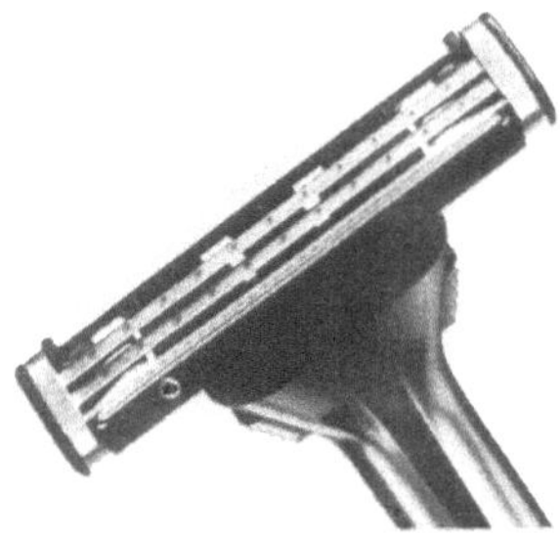

**3.3** Close-up of a Gillette Sensor razor head, showing the twin blades which are attached by precision spot welds made with an Nd:YAG laser (courtesy of Lumonics).

now be found in millions of homes. At the outset of development, the success of this floating blade concept for giving a clean, close, smooth shave depended on a precision method of joining a platinum hardened, stainless steel blade to a formed-steel support bar. For the early development work, mechanical fixing and gluing enabled successful trials to prove the shaving requirements, however, these techniques could not practically meet production demands. Experience of using $CO_2$ lasers for coil joining encouraged the Gillette engineers to conduct trials using a $CO_2$ laser to spot weld the blades to the spring assembly. These trials were so successful that a $CO_2$ laser welding system was developed for pilot production. To meet full production requirements further research indicated that Nd:YAG lasers would be more suitable due to the advantages of delivering the laser beam to the welding station via a fibre optic and the much reduced laser size would lower demand on floor space. To meet production needs 30 lasers with fibre optic beam delivery systems were initially purchased; one of the largest single orders ever placed for Nd:YAG industrial lasers. The most interesting fact about this application is that each laser makes 13 spot welds on each of the two blades affixed to the cartridge at a rate of 60 cartridges/minute, totalling 1560 spot welds; almost 100 000 spot welds per hour.

## Heart pacemakers

Implantable heart pacemakers, Fig. 3.4, are now commonplace and currently there are approximately 320 000 implants every year and, because they provide a substantial increase in life expectancy, the pro-

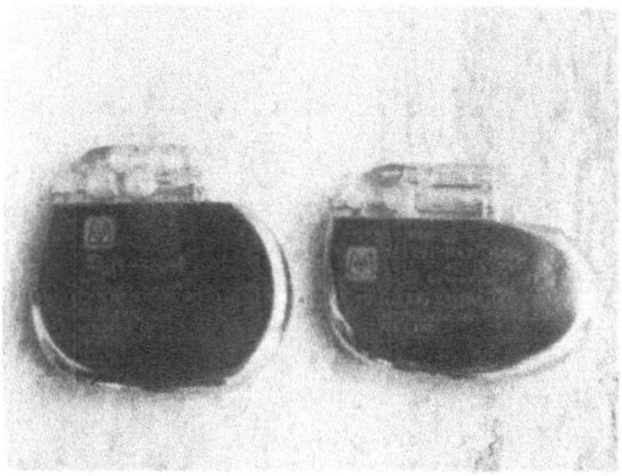

3.4 Heart pacemaker packages fabricated by an Nd:YAG laser (courtesy of Lumonics).

3.5 The continuous laser seam weld, which joins the two halves of the heart pacemaker package (courtesy of Lumonics).

duction numbers will no doubt increase. The internal components of a pacemaker comprise electronic hybrid circuits and batteries. For patient comfort the pacemaker is made as small as possible and housed in a titanium metal package fabricated from two disc like halves after the heart pacemaking electronics have been sandwiched between them. The sealing weld has to provide a very high quality hermetic seal to prevent body fluids entering the package and causing the electronic pacemaker to fail. Furthermore, the weld has to take place within 1 mm of some of the circuitry and batteries, which should not be subjected to temperatures above 50°C. Consequently a welding technique is required which can confine the weld heat to a very small zone and minimise heat transfer into the surrounding material. This objective is realised in production by using a pulsed Nd:YAG laser to make a continuous seam weld by the high speed production of a stream of overlapping spot welds, as shown in Fig. 3.5. The spot weld pulse times, of just a few milliseconds, precisely control sufficient weld energy to fuse the joint faces, without an excess of energy which could cause weld spatter and metal particles to enter the package and damage the pacemaker.

## Opto-electronic transmitters

Perhaps the major requirement for reliable and efficient operation of an opto-electronic transmitter, Fig. 3.6, is the accurate positioning and final fixing of the optical fibre in relation to the laser diode. In the example described here, the fibre of only 125 μm diameter, is held in a ferrule which has to be accurately axially aligned with the laser diode and then the position maintained for the service life by fixing the ferrule, without further movement, to the package casing. To achieve the fixing, which also has to form a hermetic seal between the ferrule and the casing, a number of joining techniques were considered, including soldering, adhesive bonding and resistance welding, however, all presented problems. Soldering is not a clean process, requires a consumable and offers problems due to creep with age or extreme temperature changes. Adhesives were ruled out as their long term stability could not be assured and also since the ferrule could move as the

**3.6** An opto-electronic transmitter, which relies on precision Nd:YAG laser spot welding to achieve the required alignment, to better than 5 μm, of its opto-electronic components.

adhesive sets. During resistance welding the electrode force caused the ferrule to move.

Laser welding, because of its low distortion of the workpiece, was examined, but initially weld shrinkage was sufficient to cause very slight movement of the ferrule in its aperture in the package casing and thus destroy its alignment with the laser diode. This problem was solved by the ability of the Nd:YAG laser beam to be split equally between three fibre optic cables and transmitted to the workpiece to perform three simultaneous spot welds, 120° apart, between the ferrule and casing. In production, using this method the opto-electric fibre is fixed and aligned to the transmitter with an alignment of better than 1 μm (laser beam splitting is described in Chapter 10). Once the ferrule is firmly positioned by the three spot welds, the workpiece is rotated through 120°, during which time a continuous, hermetic seam weld is made with a series of overlapping spot welds.

## Food mixer whisks

An early design of food mixer whisk comprised two formed, ferritic stainless steel blade strips with their ends clamped between two pressed, stainless steel discs, Fig. 3.7. The discs were then firmly resistance spot welded to ensure the blades were mechanically trapped. Increased food mixer sales demanded a review of this design and joining method in order to automate the assembly.

One of the major restrictions of the resistance spot welding operation was the fast wear of the electrodes, caused by unavoidable splashing during welding. Nickel plating of the discs to prolong electrode life was considered, but proved to be too expensive. Other joining techniques were therefore considered whilst bearing in mind a more simple piece-part design requiring the minimum material. Ultimately, a clever design transpired, Fig. 3.8, which was neat, small and self-located the assembly prior to welding. The success of the design depended on finding a high speed joining technique which could join all three component parts simultaneously, limit metallurgical damage to the surrounding material and have a good appearance. After an examination[2] of hard soldering and fusion welding techniques, only electron beam and laser welding (two power beam techniques) were found to meet these requirements. Laser welding was subsequently chosen in preference to electron beam welding, because the necessary vacuum chamber for

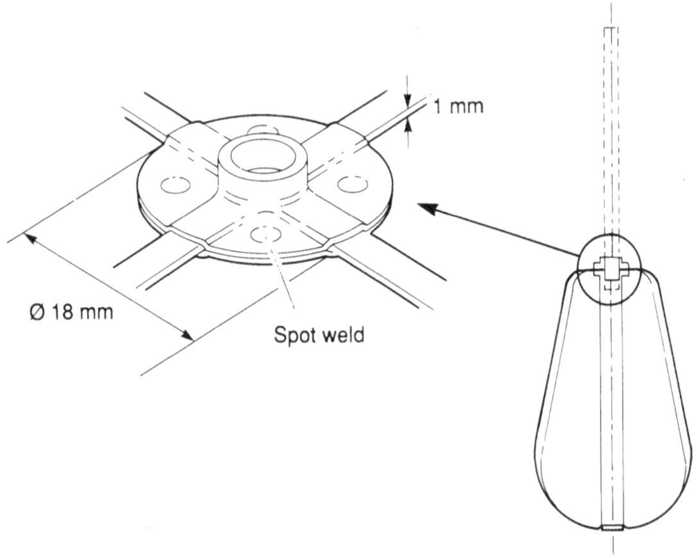


3.7 The original design for joining food mixer whisk blades by resistance spot welding.

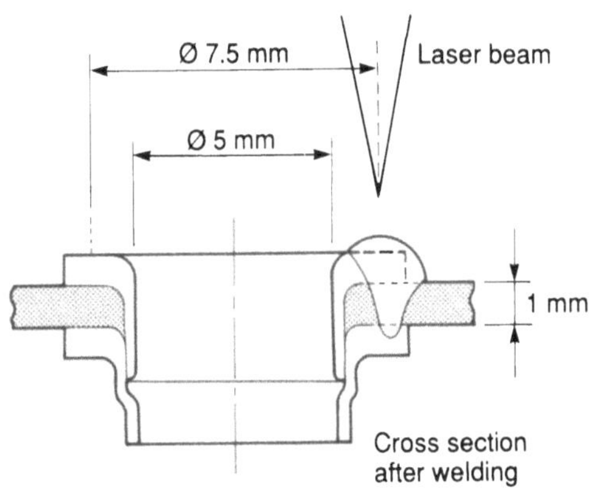


3.8 The self-locating design adopted for joining food mixer whisk blades during automated $CO_2$ laser welding.

electron beam welding made the automatic workhandling equipment much more complex than for laser welding.

Production welding of the revised design whisks is conducted with a $CO_2$ laser, which automatically welds over one thousand whisks per hour.

3.9 Laser welded hydraulic tappet housings (courtesy of Daimler Benz).

## Tappet housing

The tappet housing described here is an integral part of a hydraulic valve-play compensator and used in a passenger car diesel engine. It comprises two, pressed steel, pot shaped components, which are pressed into one another as shown in Fig. 3.9. The external diameter of the assembly is just over 40 mm and the wall thickness in the joint area, which is halfway down the inside of the assembly, is just over 1 mm. The materials are low alloy case-hardening steel (16MnCr 5/15 Cr3). The weld is required to fix firmly the relationship of the two pressings and be airtight.

During the development stage, brazing, electron beam and laser welding were examined for joining the two pressed parts. Brazing produced a satisfactory joint, but created problems due to the braze material obstructing oil ducts necessary for the function of the component. Electron beam and laser welding both produced very satisfactory welds and for the required weld cross section, Fig. 3.10, the welding operation speeds were approximately the same. However, since the laser welding did not require a vacuum chamber, handling the components before and after welding was considerably simpler and the equipment size kept substantially smaller. Moreover, equipment maintenance costs for laser welding were calculated[3] to be lower than for electron beam welding.

A 1.5 kW $CO_2$ laser is used in the automated production welding of the tappet housing, producing a weld depth of $0.7^{+0.3}_{-0}$ mm at a welding speed of 2.7 m/min. These permit a cycle time of 4 seconds, which can equal a yield of 8000 components in two shifts.

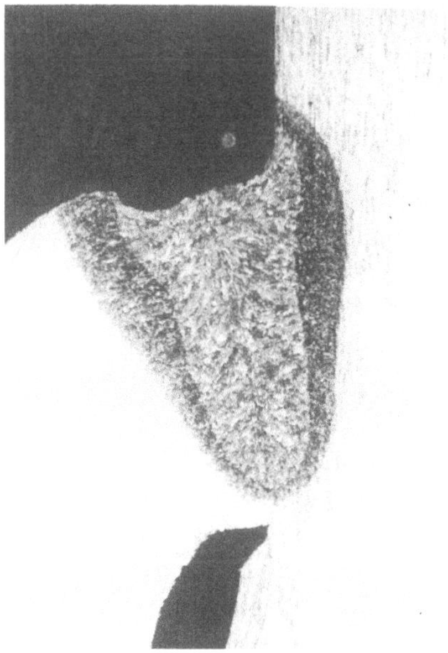

3.10 Macro-section of weld joining the hydraulic tappet housing shown in Fig. 3.9 (courtesy of Daimler Benz).

## Gas hob distribution plates

For many years, the method transmitting gas from domestic cooker taps to the burners on the hob has been either through cast manifolds or an assembly of pipes, elbows and unions. In the early 1980s, New World designed an innovative manifold system for a four burner hob based on two pressed, aluminised, mild steel sheets. The success of this simple and cost effective design depended on finding a reliable, fast and efficient method of joining the two sheets. The four gas routes had to be individually gas tight and part distortion, such as buckling, kept to a minimum.

After careful consideration the choice of techniques for fabricating the manifold was narrowed down to adhesive bonding and laser welding. The manifold design incorporated a lap seam weld, Fig. 3.11, which is the easiest sheet metal joint to make by laser welding (Chapter 6) assuming the parts are clamped firmly together and the laser beam focus position is carefully maintained. At a welding speed of 3 m/min and using a 2 kW $CO_2$ laser, distortion-free welds of the required quality were made with a good cosmetic appearance. Also, as the weld fusion

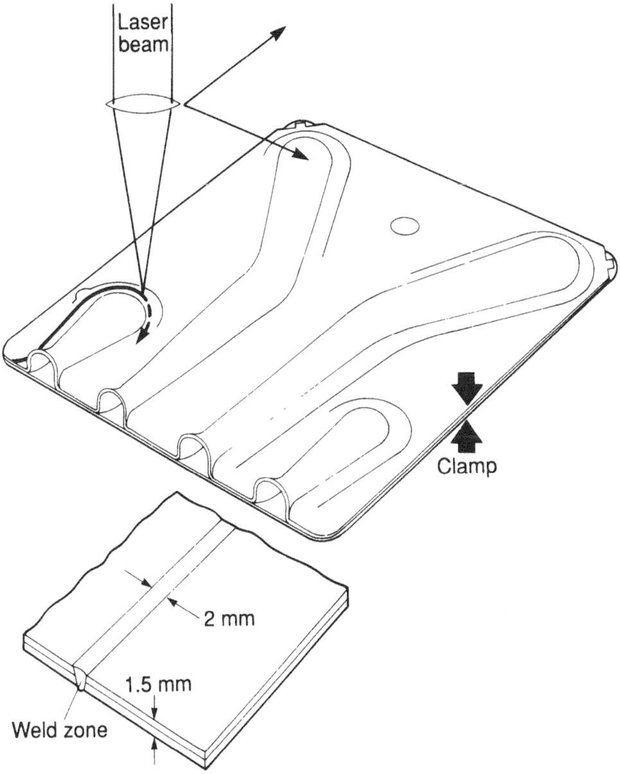


3.11 The joint configuration details and workpiece/laser beam manipulation concept used to lap seam weld the gas hob manifold.

zone penetrated the lower sheet, visual inspection could confirm the weld had adequately penetrated across the joint interface. On the other hand, the trials with adhesive bonding were also successful, however, the choice of joining technique was finally resolved in favour of laser welding, because of the higher level of long term confidence in the welded joint.

Rather than buy a laser welding system for immediate in-house production of the gas hob distribution plate, New World, who had no previous production experience of laser welding, wisely chose to have their early production volumes welded by a laser welding job shop. This approach allowed New World to gain confidence before committing a large capital sum on equipment and staff training. Furthermore, early problems could be quickly solved by experienced laser people.

The job shop produced over 200 000 gas hob distribution plates between 1985 and 1988. For the welding operation, they clamped the plate assembly in a pneumatic press and then, under computer control moved the laser gun around the required weld path on the stationary plate assembly.

In 1988, New World purchased their own laser welding system, which deviates from the job shop's in that the plate assembly is moved beneath a stationary laser gun. (Laser beam and workpiece manipulation systems are described in Chapter 10.) With this system, New World produces 450 welded assemblies per day, each being subsequently pressure tested. The failure rate is extremely low and those that fail can usually be reclaimed. When the laser welding operation was brought in-house, the cost was reduced by 47%. In 1988 the cost of the laser system was £115 000 and the payback was covered in 1.9 years.

## Passenger car body parts

The neat, strong, low distortion laser weld has made a significant contribution in the quest for lighter, corrosion-free car bodies and a reduction in material wastage when blanking body parts. Three examples which between them embrace these points are described here.

The first example is a car floor pan. In this case the car body manufacturer wished to make a single piece floor pan from a large galvanized sheet 1960 × 3200 × 0.75 mm. Since there were no production installations anywhere in the world producing such sheet sizes in galvanized steel, ways of manufacturing sheets were investigated by Thyssen Stahl of West Germany. The obvious approach was to join two sheets to produce one of the required size.

To join the galvanized steel sheets a welding technique was required which would produce a weld capable of meeting the following requirements:

- Sufficient ductility to allow blanking and pressing operations;
- Weld top and bottom beads flush with the sheet surface to prevent blanking and press tool damage;
- A very narrow weld to keep the heat affected zone to a minimum and preserve the cathodic fusion effect of the zinc.

Initial welding trials, which examined TIG welding, foil butt seam welding and mash seam welding, did not meet the above requirements. At the time laser welding was just industrially established for welding engine and transmission components, but nobody had laser butt welded 0.75 mm thick sheets with weld lengths almost two metres long.

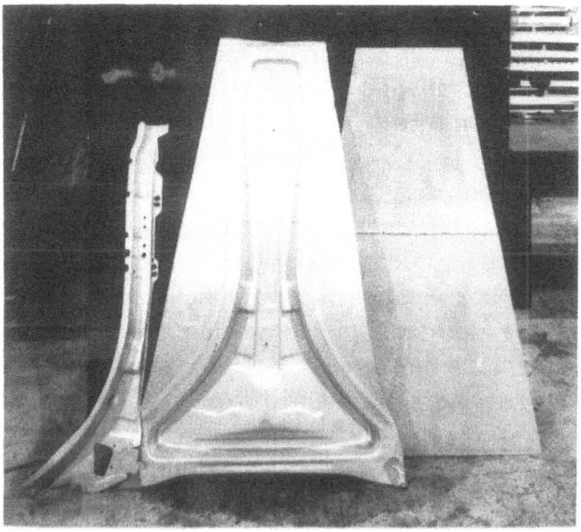

**3.12** Stages of producing a motor compartment rail that has a change in material thickness, which is achieved by fabricating the original blank by joining a 0.8 mm thick sheet to a 2 mm thick sheet by a laser weld (courtesy of Utilase).

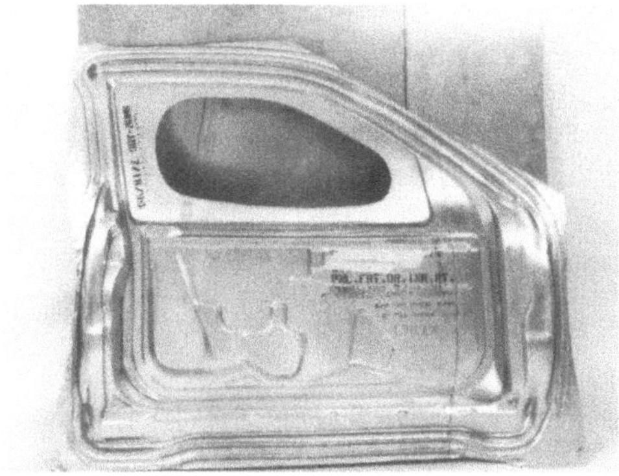

**3.13** A similar situation to that described in Fig. 3.12, but concerning an inner door panel. The laser weld can just be seen as a fine line running vertically down towards the righthand side of the panel. A pre-fabricated door blank stands behind the pressing (courtesy of Utilase).

Nevertheless, Thyssen Stahl recognised the potential advantages of laser welding and through careful attention to detail and precision control of the welding procedure[4] developed and launched a successful large scale production facility, producing some 40 000 laser welded sheets per month.

For the second example, Fig. 3.12 shows a motor compartment rail, made from deep drawing grade low carbon steel, in three stages of its manufacture. Part of the rail has to be of heavy duty gauge (2 mm) steel for adequate seat belt component anchorage strength. Rather than press the complete rail from 2 mm thick sheet, a pre-pressed blank is fabricated by $CO_2$ laser butt welding a 0.8 mm sheet, representing the zone where high strength is not required, to a 2 mm sheet thus reducing material cost and weight. In Fig. 3.12 the top of the unpressed panel is the 0.8 mm thick steel and the lower half the 2 mm one. The post-weld forming operation is a stringent test of weld quality and a weld after forming can be seen on the pressed panel in the centre of the picture. From this application it will be realised that laser welding can reduce material wastage, as such blanks can be fabricated by joining scrap parts left over from blanking operations carried out elsewhere.

A similar example to that just described is shown in Fig. 3.13. Here, material and weight has been saved on an inner door panel by joining a 1.8 mm thick sheet, in a position where strength for hinge and mirror assemblies is required, to a 0.8 mm thick sheet which does not require the same mechanical strength. A laser butt welded sheet assembly prior to blanking and forming is standing behind the inner door panel and on which the laser weld can be seen as a fine line running vertically down the pressing.

## Seam welded stainless steel tubes

High frequency (HF) induction welding has little competition from any arc or power beam technique for manufacturing seam welded thin wall, mild steel tubes; especially where high production rates are required. Unfortunately, when this technique is applied to the manufacture of stainless steel tubes, the weld quality is often found to be poor when the tubes are subjected to cold forming; such as simple bending. When welding mild steel the high squeeze roll forces, applied to the induction heated faying surfaces, are able to break up and squeeze out most of the oxides and contaminates that would otherwise prevent acceptable weld formation. In the case of stainless steel, it is difficult during HF induction heating to eliminate the formation of detrimental chromium and iron oxide films at the joint interface and then remove them in the squeeze rolling operation.

It is possible to make acceptable seam welds in stainless steel by arc welding techniques such as TIG and plasma. However, welding speeds are slow and great care is necessary to prevent overheating the stainless steel, otherwise metallurgical damage occurs which reduces the material corrosion resistance (Chapter 4).

Laser welding cannot meet the welding speeds of HF induction welding, but it is generally at least three times faster than mechanised TIG welding and the very narrow keyhole welds and narrow heat-affected zones achievable can considerably restrict potential metallurgical damage in stainless steel. Moreover, weld reject rates are considerably lower than when HF induction welding stainless steel. The most significant factor is perhaps the good cold formability of laser welds, which is illustrated in Fig. 3.14. These facts have led several seam welded stainless steel tube manufacturers to turn to laser welding.

**3.14** An example of the formability of a laser seam welded stainless steel tube (courtesy of Rofin-Sinar).

## Stainless steel heat exchangers

In heat exchanger applications, stainless steel finned tubes are often required for the optimum design for a specific application. Finned tubes are generally produced by helically winding strip on to tubes, Fig. 3.15. The heat exchange efficiency of the finning depends on good heat conduction across the joint between the fin and tube interfaces.

The ideal solution is a continuous metallic bond across the entire joint interface and therefore a joining technique which can place such a joint in this position is required. Because the corrosion and strength properties of stainless steel are impaired if held at elevated welding temperatures for a long period, a high speed keyhole welding technique or a much lower joining temperature hard soldering or vacuum brazing technique is required. In this instance, hard soldering and brazing techniques require more expensive consumables and can be difficult to apply to long lengths of finned tube, which can be up to 10 m long.

The heat exchanger fin to tube joint is a T joint configuration and from the advantageous characteristics of laser welding previously described in Chapter 2 and reference to Fig. 2.4, one no doubt has already concluded that laser welding is ideally suited to this application. This particular advantage has been grasped for some time by KM-Schmöle GmbH who manufacture such tubing and market it under the name 'Laserfin Tube'. Cross sections through T joints in thin sheet steel shown in Chapter 6 nicely illustrate the confinement of the weld zone close to the original joint interface. The sections also show the confinement of the heat affected zone.

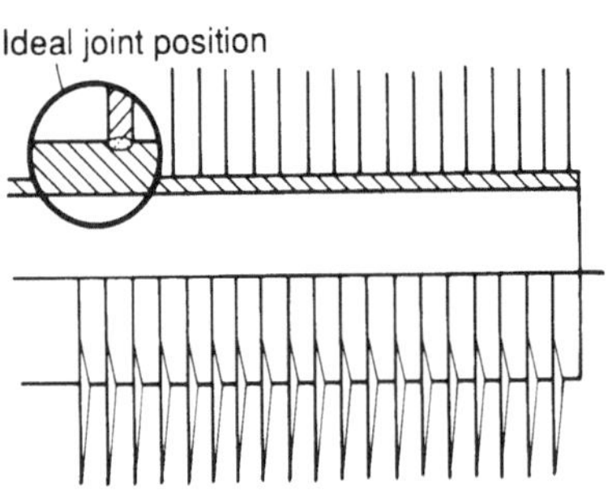


3.15 The principle of helically wound finned heat exchanger tubing and the ideal fin to tube joint position.

## Saw blades

The joining of diamond or carbide impregnated steel tips to rotary machine tool saw blades by laser welding is emerging under certain circumstances, as a viable and cost effective alternative to brazing.[5] Providing the impregnated particles on the tips can be kept away from the weld zone and the disc, to which the tips are attached, is of a suitable material for producing a weld which can withstand the service requirements, then laser welding has numerous advantages over brazing. The operation is quicker and cleaner, process consumables are less expensive, post-joining machining is unnecessary, and part distortion is reduced. Furthermore, in service, the laser welded tips do not require water cooling to prevent the brazed joint melting and causing tip failure.

## Solenoid operated hydraulic valves

A major application of laser welding is the fabrication of shuttle valve cylinder assemblies, Fig. 3.16, for solenoid operated hydraulic valves. There are several designs which use laser welding in their fabrication and in each case laser welding has been chosen for the same reasons. Apart from a repeatable, good quality, high strength weld, the main reasons, which initially attracted attention to laser welding, were the need to butt weld, with good alignment and low distortion, the non-magnetic stainless steel and mild steel tubular parts which form the

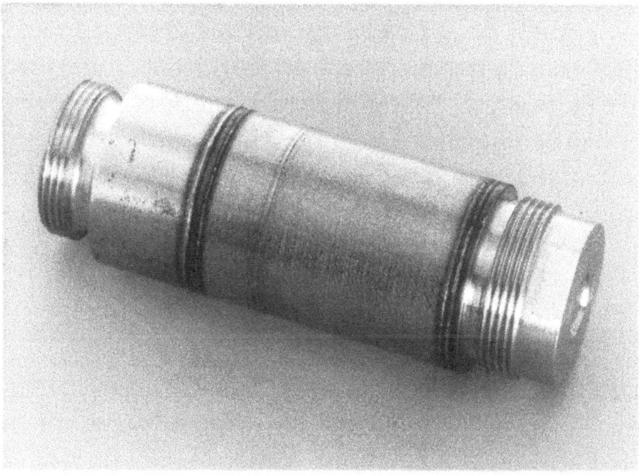

**3.16** A pre-machined solenoid valve and cylinder assembly which is completely assembled and then finally sealed by two laser welds. The diameter of the assembly shown is approximately 20 mm (courtesy of Vickers Systems).

main valve cylinder without damaging internal rubber O rings during final assembly, plus minimising the need for post-weld machining and unnecessary material waste.

Early trials using arc welding caused excessive distortion and unacceptable internal weld beads, both of which would have necessitated making the parts oversize demanding substantial post-weld machining. Friction welding, which can give a very high quality weld, was also tried by one manufacturer, but discontinued on small valves due to material wastage through axial shortening during welding and the extra machining time for removing the internal and external weld flashes.

Trials with laser and electron beam welding showed at welding speeds between 2 and 4 m/min, the part distortion was very low and near to the tolerance required for the shuttle valve operation. On early production valves the weld between the stainless steel and the mild steel produced an underbead which penetrated into the valve cylinder bore. This together with bore eccentricity was quickly removed by a reaming operation. Owing to the low distortion achieved, it became apparent that if the cylinder parts were accurately, concentrically pre-assembled with the aid of a small spigot, and the weld penetration controlled to stop just at the cylinder bore, the material removal during the reaming operation could be substantially reduced. Having proved this point, it was a short step to manufacture valve assemblies, as shown in Fig. 3.16, whereby the pre-assembled shuttle valve and cylinder assembly is also finally sealed by two laser welds, employing spigotted joints, which also permit a design which eliminates screwed in end caps.

Today $CO_2$ lasers are generally used worldwide, for fabricating the type of valve in question. Automatic work handling equipment is used on a range of different valve assemblies, with welding cycle times typically from 8 to 16 seconds per assembly. One UK manufacturer has claimed production rates of 50–60 000 assemblies per month.

## Viscous couplings and differentials

Viscous couplings and differentials are components used in four wheel drive systems. A typical unit consists of an outer housing with an internal spline form, Fig. 3.17, and inner hub with an external spline form, and two types of plate. Alternate plates locate on the inner and outer spline forms and there is a viscous medium between them which transmits the drive.

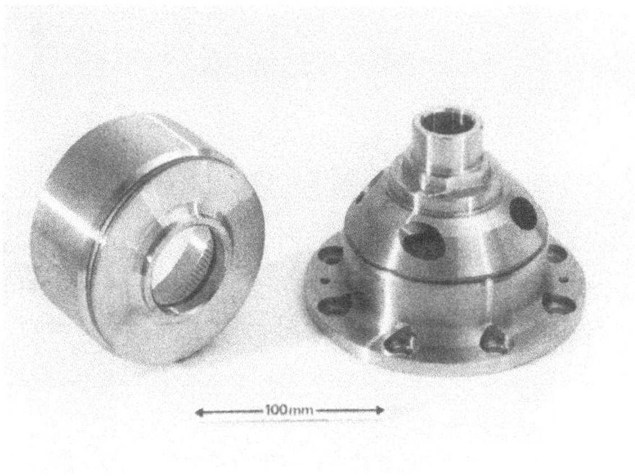


**3.17** Laser welded, pre-machined viscous drive components used in four wheel drive vehicles (courtesy of GKN/Uni-Cardan).

The internal splines made manufacture of the outer housing from a single part difficult, because through-broaching to give a quick and precise spline form is not possible. Consequently, a construction comprising two pre-machined parts is used. Electron beam and laser welding are the only two fusion welding techniques which can join these pre-machined parts and maintain the required dimensional accuracy after welding. The parts could of course be bolted together, but this approach would require extra materials and machining operations. In production the parts, in a 0.3 carbon steel, are joined by a 3 mm deep circumferential butt joint made with a laser power of 5 kW and welding speed of 1.8 m/min. To meet production numbers a single laser works several welding stations in a similar manner to that shown in Chapter 2, Fig. 2.9.

## Specialist fabrication

The examples of laser welding applications described so far all concern components manufactured in substantial numbers. There are, however, many one-off applications which are handled by laser welding specialists. The example shown in Fig. 3.18 is part of a hydraulic bearing thrust unit and shows that lasers can be used for welding large components. The material is 10 mm thick type 321 stainless steel. The assembly is approximately 2.5 m in diameter and 0.7 m deep and comprises four laser welds. The first two welds butt join two rolled semi-circular sheets

**3.18** A hydraulic bearing thrust unit 2.5 m diameter in type 321 stainless steel, 10 mm thick, being fabricated with a 10 kW laser.

to form the tubular section. The final two welds join L section rings, via butt joints to the tube ends, thus forming flanges. The welding operation, which is shown in progress, was conducted with a 10 kW laser and at a welding speed in the order of 1.8 m/min. Arc welding was considered for this application but a preliminary examination indicated the necessary multipass welding operation would cause excessive dimensional distortion through weld shrinkage (18 mm on diameter). After laser welding the final diameter shrinkage of the assembly was approximately 6 mm.

## References


1 Notenboom G J A M: 'Laser spot welding in the electronics industry'. Seminar on 'Practical application of lasers in manufacturing industry', Coventry, The Welding Institute, Nov 1982.
2 Notenboom G J A M and Jelmorini G: 'High power $CO_2$ laser welding of food mixer parts, a case study'. Proc of SPIE Innsbruck Conf 1986.
3 Luttke H: 'Use of laser and electron beams in seam welding of gear and engine parts'. Seminar on 'Applications of laser processing in automobile fabrication and related industries', Cambridge, The Welding Institute, Dec 1987.
4 Prange W: 'Use of laser welding technology in sheet processing'. Seminar on 'Applications of laser processing in automobile fabrication and related industries'.
5 Havrilla D: 'Laser welding of saw blades'. Rofin-Sinar Inc Special Publication, March 1990.

# Chapter 4

# Materials

This chapter outlines materials which can be laser welded; the effects of alloying elements and surface coatings on the weldability of common engineering materials; some weld properties data; and finally basic notes on possible metallurgical defects in welds and how to avoid them.

Laser welding is normally a liquid-phase (fusion) welding process, i.e. it joins metals by melting their interfaces and causing the mixing of

***Table 4.1* Fusion weldability of pure metal combinations**

| | Silver | Aluminium | Gold | Beryllium | Cobalt | Copper | Iron | Magnesium | Molybdenum | Nickel | Platinum | Rhenium | Tin | Titanium | Tantalum | Tungsten |
|---|---|---|---|---|---|---|---|---|---|---|---|---|---|---|---|---|
| Aluminium | B | | | | | | | | | | | | | | | |
| Gold | A | E | | | | | | | | | | | | | | |
| Beryllium | E | B | E | | | | | | | | | | | | | |
| Cobalt | C | E | B | E | | | | | | | | | | | | |
| Copper | B | B | A | E | B | | | | | | | | | | | |
| Iron | C | E | B | E | B | B | | | | | | | | | | |
| Magnesium | E | B | E | E | E | E | C | | | | | | | | | |
| Molybdenum | C | E | B | E | E | C | B | C | | | | | | | | |
| Nickel | B | E | A | E | A | A | B | E | E | | | | | | | |
| Platinum | A | E | A | E | A | A | A | E | B | A | | | | | | |
| Rhenium | C | D | D | E | A | C | E | D | E | C | B | | | | | |
| Tin | B | B | E | C | E | B | E | E | C | E | E | C | | | | |
| Tantalum | E | E | D | E | E | C | E | D | A | E | E | E | E | | | |
| Titanium | B | E | E | E | E | E | E | C | A | E | E | E | E | A | | |
| Tungsten | C | E | D | E | E | C | E | C | A | E | A | E | C | A | B | |
| Zirconium | E | E | E | E | E | E | E | C | E | E | E | E | E | B | A | E |

Note:
Those combinations which have poor thermal conductivity will probably be good absorbers of laser light. Good thermal conductors will be more suited to short wave length light such as from Nd:YAG lasers.

A — Good (solid solubility achieved)
B — Acceptable (doubtful weld structures are possible)
C — Weld with caution (unacceptable data available)
D — Take extreme caution (no data available)
E — Undesirable combination (intermetallic compounds formed)

the molten metal which solidifies on removal of the laser heat source. (Metals can be welded without melting by solid-phase welding techniques such as friction or cold pressure welding.[1] A laser could in theory be used as a heat source for making solid-phase welds, however, no industrial applications are known to the author.) Lasers can be used for soldering and brazing; these applications are outside the scope of this book.

The most suitable metals for laser welding are those which have the same or overlapping liquid temperature ranges and are soluble in one another; assuming of course they are good absorbers of laser light. Pure metals which have one definite melting temperature can usually be readily fusion welded to themselves, but not always to a different pure metal, Table 4.1. They will, however, usually weld to an alloy which has the same base metal element. Table 4.2 gives a general guide to numerous engineering materials which can be laser welded.

***Table 4.2* A guide to engineering materials which can be laser welded**

| *Material* | *Comments* |
|---|---|
| Aluminium alloys | Laser welds in aluminium alloys generally have poor quality and caution is necessary for applications where high reliability is required. See text for further information. |
| Copper | Only suitable for micro-spot welding applications, see text. |
| Cast iron | Nodular cast iron can be welded using a nickel filler wire to overcome weld cracking. |
| Nickel base and high nickel alloys (C263, Hastelloy, Inconel, Incoloy, Kovar, Monel, Nimonic, PK33, Waspaloy) | Some of these alloys weld extremely well, however, those without a number are available in different grades which produce quite different weld qualities. Therefore, the material manufacturer should be consulted with respect to weld properties, in case a filler material is required to improve them. |
| *Steels* | |
| Low carbon and high strength low alloy (HSLA) forming grades | Very good quality welds can be achieved provided sulphur and phosphorus levels are kept low, see text. |
| Medium and high carbon | Weldable, but special precautions are necessary to ensure acceptable weld properties, see text. |

*Table 4.2 cont'd*

| | |
|---|---|
| Alloy steels | Satisfactory laser welds have been made in numerous pipeline, shipbuilding and structural steels, see Table 4.3. High weld hardness can be a problem due to fast cooling rate. |
| *Stainless steels* | |
| Austenitic | Very good quality welds can be achieved in most grades except free machining, see text. |
| Ferritic | Grades with low carbon and chromium levels weld best. Weld toughness is affected by grain coarsening, see text. |
| Martensitic | Welds and their HAZs are hard and brittle due to the high carbon content, see text. |
| Titanium and alloy 6Al-4V-Ti | Good quality welds with fine grain structures can be achieved, but material cleaning just prior to welding and high quality weld pool gas shielding are essential. |

## Weldability and alloying elements

The first thoughts when choosing a material are usually about whether it will fulfill its service and manufacturing requirements, i.e. strength, toughness, wear resistance, formability, machinability, etc. To meet the service and manufacturing requirements the chosen material will contain numerous alloying elements, some of which can affect weld quality. Because a material can be welded (by whatever process) it does not necessarily mean that the welds are always mechanically and metallurgically sound. Weld properties, weld metal structure, strength and toughness, etc, depend on the parent metal composition and the resultant heat treatment which occurs due to the welding operation. The heat treatment causes changes in the structure of the weld metal and HAZ and in certain material compositions unacceptable metallurgical damage can occur; certain steels and most aluminium alloys are susceptible. Fortunately, manufacturers provide information as to the weldability of their materials and specific grades. Furthermore, they will provide a chemical analysis of the alloying elements present. As it is not possible to present manufacturers' data here, the weldability and some of the elements which can affect weld quality in some of the common engineering materials will be described. More detailed attention will be given to steel as it is the most used material.

## Steel

A general guide to the weldability of common types of steel is given in Table 4.2 and data on the mechanical properties of laser welds which have been made in pipeline, shipbuilding and structural steels is given in Table 4.3.

***Table 4.3*** **Mechanical properties of various laser welds made in pipeline, shipbuilding and structural steels**

| Steel type | Composition | Thickness, mm | Power, kW | Speed, m/min |
|---|---|---|---|---|
| X-80 arctic pipeline | 0.06C, 1.35Mn, 0.002S 0.6Si, 0.065Al, 0.1Nb 0.025Cr | 13.2 | 12 | 0.64–0.76<br>1.5–1.65 |
| HY-130 ship-building (quenched and tempered) | 0.1C, 0.25Mn, 0.5Cr 0.4Mo, 0.07V, 5.5Ni | 6.4 | 5 | 0.5–1.14 |
| X-60 pipeline | Various, but typically: 0.15C, 1.35Mn, 0.0015S 0.1Si, 0.015Al, 0.03Nb | 12 | 10–15 | 0.7–1.32<br>1.0–1.77 |
| | | 18 | 12<br>14 | 1.14<br>1.14 |
| | | 26 | 12<br>14 | 0.76<br>0.88 |
| HY-130 shipbuilding (quenched and tempered) | 0.11C, 0.80Mn, 0.6Cr 0.54Mo, 0.07V, 4.7Ni | 6.35<br>12.7 | 8–12<br>8–12 | 1.38–1.5<br>0.76–1.0 |
| BS 1501: 271 Grade W30 | 0.12C, 0.29Si, 1.49Mn 0.24Ni, 0.46Cr, 0.25Mo 0.11V, 0.01Nb, 0.025S 0.026P | 6 | 5 | 0.96 |
| HY-130 shipbuilding (quenched and tempered) | 0.11C, 0.55Cr, 0.55Mo 5.0Ni, 0.8Mn | 12 | 11 | 0.75 |
| HY-80 shipbuilding (quenched and tempered) | 0.18C, 1.68Cr, 0.45Mo | 12 | 10.6 | 0.75 |
| Grades A, B, C shipbuilding | Various, but in the ranges: 0.23 max.C, 0.6–1.03Mn | 9.5–28.6 | 5.5–12.8 | |
| BS 4360 Grade 50D | C-Mn-Si-Al-Nb Various | 12.5 | 5.2–9 | 0.18–1.5 |

Steel is a good absorber of the light wave lengths produced by $CO_2$ and Nd:YAG lasers and many steels are readily weldable. However, weld formation and quality are considerably influenced by the following elements when present in unacceptable levels.

| Single/double pass | Weld transverse tensile tests | Toughness tests | Reference |
|---|---|---|---|
| Single<br>Double | All failed in base metal | Standard Charpy. Upper shelf energy >358J. Transition at −9°C for single pass welds, −51°C for dual pass | 2 |
| Single | All failed in base metal | Half size Charpy. Weld metal absorbed more energy than the base metal at several temperatures | 3 and 4 |
| Single<br>Double<br>Double<br>Double<br>Double<br>Double | Failed in base plate or at same UTS if failed in weld<br>Failed in base material | Standard Charpy. Upper shelf energy often higher than base material. Transition at ~−20°C | 5 |
| Single<br>Single | Welds produced at optimum conditions failed in base material | Subsize dynamic tear tests. Fractured in the base material at room temperature, absorbed energy 518J (6.35mm plate). 1030J (12.7mm plate) | 6 |
| Single | All failed in base material | Not tested | 7 |
| Single | Four out of 16 tested failed at weld. Porosity visible on fracture surface | Subsize dynamic tear tests. Transition at −18°C, 707J absorbed energy at room temperature | 8 and 9 |
| Single | — | Subsize dynamic tear tests. 76J absorbed energy at room temperature | |
| Both | All failed in base material | Standard Charpy. Absorbed energies at −15°C in range 11–49J | 10 |
| Single | — | Standard Charpy. 27J at −40°C achieved. | 11 |

*Oxygen*

Oxygen is an element which is used in the making of steel and if allowed to remain in the steel in high levels, e.g., more than 100 ppm tends to outgas significantly (especially in thin sheet steel) during fusion welding promoting porosity and sometimes cracks in the weld metal. Therefore, a killed steel (a steel where the oxygen content has been reduced to a low level) and not a rimmed steel should be used for laser welding. Moreover, a killed steel has a more uniform composition which enables a more consistent weld quality to be achieved.

*Carbon*

Carbon is alloyed with iron to increase hardness and strength and the level of carbon is dependent on the potential application for the steel. As the level of carbon in different steels increases, so do the potential problems which can adversely affect weld quality and consequently service requirements. The low carbon steels which have carbon levels of less than 0.1% produce good quality welds and reliable weld performance; unless of course other undesirable elements are present. The narrow laser welds which are produced in low carbon steels are quite hard. This is because the rapid welding speed causes high cooling rates which enable limited transformation hardening (heat treatment) to take place. Figure 4.1 shows the parent metal, HAZ and weld metal hardness in a laser welded lap joint between two sheets of 0.7 mm thick, 0.04% carbon steel after welding at 6.5 m/min and using a $CO_2$ laser power of 5 kW. The hardness of the parent metal is less than 150 Hv, whereas the weld metal is more than twice this figure. Yet, surprisingly, such narrow and hard welds, when made in deep drawing low carbon grades of steel, have good ductility. This can be seen in Fig. 4.2 which shows a laser welded butt joint in a low carbon steel hoop, 240 mm diameter, before and after forming into a 280 mm diameter wheel rim. The welded joint in this case has withstood a rigorous forming operation.

In steels where the carbon content is increased beyond 0.1%, the rapid weld cooling rate produces a martensitic microstructure, a hard and brittle phase, which causes a reduction in toughness and ductility. The very hard martensite can cause cold cracking in both the weld metal and HAZ. This is a brittle metal crack which propagates during the later stages or after cooling and will be discussed shortly.

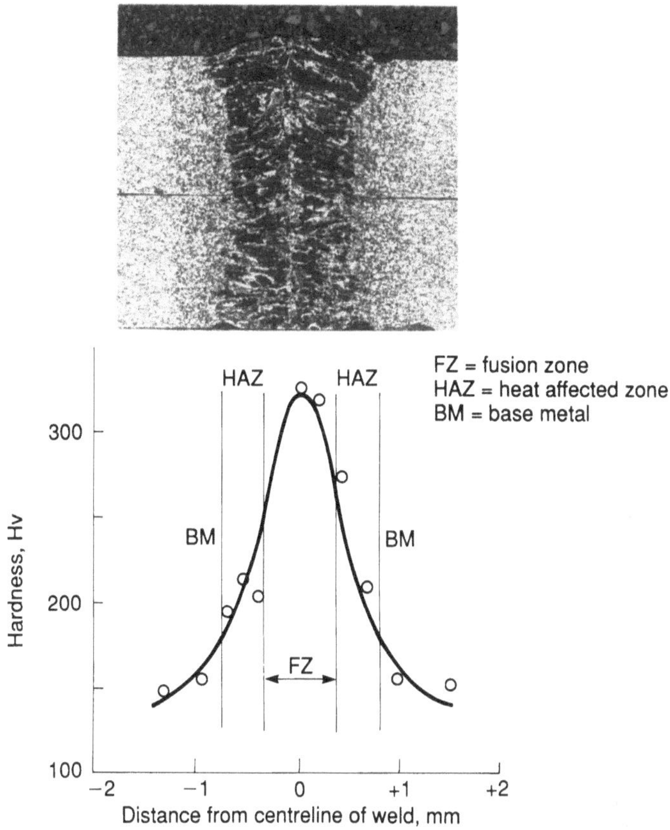


**4.1** Section through a lap seam welded joint between two 0.7 mm thick, 0.04% carbon steel sheets and a hardness plot taken across the weld. This illustrates the high weld hardness produced when laser welding a relatively low carbon steel.

Despite the potential drawbacks when welding steels with carbon levels greater than 0.1%, laser welds are made in certain gear steels with carbon levels up to 0.3%; as shown in the previous chapter. However, careful design and welding procedures can be adopted (Chapter 7) to eliminate cracking and to ensure that the welds will meet their service requirements.

The carbon content in steel, although the major element concerned, does not alone determine the weld hardness for a given cooling rate. This is decided by the presence of other alloying elements such as manganese, chromium, molybdenum, vanadium and copper. To establish whether the percentage of these elements when present in a low alloy steel is likely to cause problems associated with high weld

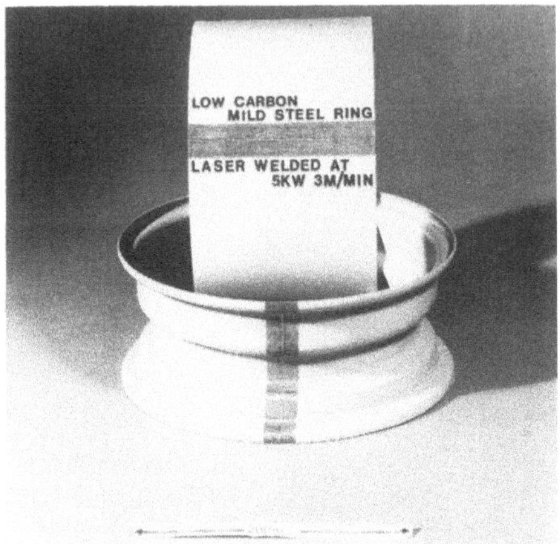


**4.2** A laser welded butt joint in a low carbon steel hoop, 240 mm diameter, before and after forming into a 280 mm diameter wheel rim. This illustrates the weld ductility despite the hard weld metal.

hardness, such as solidification and HAZ cracking, numerous carbon equivalent (CE) formulae have been developed. The British Standards Institution's CE formula is as follows:

$$CE = C\% + \frac{Mn\%}{6} + \frac{(Cr\% + Mo\% + V\%)}{5} + \frac{(Ni\% + Cu\%)}{15}$$

A CE less than 0.40 indicates the steel is readily weldable, but above this figure precautions such as careful control of weld heat input or pre-heating may be necessary to eliminate possible weld and HAZ faults.

*Sulphur and phosphorus*

Sulphur and phosphorus are present in steel as residual elements from steelmaking and their levels are controlled according to the processing requirements of the particular type of steel. Sulphur assists machinability and therefore sufficient levels for this are left in free cutting steels. Unfortunately, very small amounts of sulphur and phosphorus can promote serious weld solidification cracking and therefore their combined level is kept as low as possible in weldable grades of steel; free cutting steels are therefore most unsuitable for welding. When welding joints with low mechanical restraint, e.g. full penetration butt welds in small components where the joint faces can be forced together during the welding cycle, combined sulphur and phosphorus levels of 0.04% can be tolerated without promoting cracking. However, when welding joints with high mechanical restraint e.g. partial penetration butt welds in thick sections, combined levels of no more than 0.02% are recommended to avoid weld cracking.

Solidification cracking, which will also be discussed shortly, occurs under the influence of high weld contraction stresses during rapid weld solidification and cooling. The contraction stresses in partial penetration welds in thick sections can be reduced in some instances by attention to joint design and is discussed in Chapter 7.

## Stainless steel

Apart from the special alloys, stainless steels fall into three basic types: austenitic, ferritic and martensitic. All three can be laser welded, but the weld quality of autogenous welds (welds made without the addition of a filler material, which can be selected to modify the weld metal) decreases in the order which the three types are presented and described below.

### *Austenitic stainless steel*

Austenitic stainless steel, which is usually referred to as the 300 series, is ideally suited to laser welding, Table 4.4, with the exception of grades 303 and 303Se which contain added sulphur and selenium; these elements, which aid free machining can produce hot cracking. Austenitic stainless steel has a thermal conductivity of one third of carbon steel and is a superior absorber of laser light. Consequently, slightly higher weld penetration depths can be achieved than with carbon steels for a given welding condition.[12] Besides ease of welding, one of the reasons austenitic grades are ideally suited to laser welding is because the low heat input and high welding speeds do not allow metallurgical damage to occur which can impair the corrosion resistance of the weld metal. If austenitic stainless steels are held for prolonged periods in the 450 to 870°C range, the chromium-rich carbides begin to precipitate at the grain boundaries and this reduces the corrosion resistance; a common occurrence when using high general heat input welding techniques.

Another advantage of using a laser to weld austenitic stainless steels is the low weld distortion produced by the low general heat input and the laser weld shape. The austenitic grades have a 50% greater thermal

*Table 4.4* **Austenitic steels (common grades)**

| Specifications | | | | | | | |
|---|---|---|---|---|---|---|---|
| AISI | BS 1449 | Others | C | Cr | Ni | Mo | Others |
| 304 | 304 S16 | W.Nr. 1.4301 | 0.06 max | 17.5–19.0 | 9.0–11.0 | | |
| Typical | | | 0.05 | 18.4 | 9.5 | | |
| 304L | 304 S12 | W.Nr. 1.4306 | 0.03 max | 17.5–19.0 | 9.0–12.0 | | |
| Typical | | | 0.02 | 18.4 | 9.3 | | |
| 309 | 309 S24 | | 0.15 max | 22.0–25.0 | 13.0–16.0 | | |
| Typical | | | 0.06 | 23.0 | 14.3 | | |
| 310 | 310 S24 | W.Nr. 1.4845 | 0.15 max | 23.0–26.0 | 19.0–22.0 | | |
| Typical | | | 0.10 | 25.0 | 20.0 | | |
| 316 | 316 S16 | W.Nr. 1.4401 | 0.07 max | 16.5–18.5 | 10.0–13.0 | 2.25–3.0 | |
| Typical | | | 0.055 | 17.0 | 11.9 | 2.5 | |
| 316L | 316 S12 | W.Nr. 1.4404 | 0.03 max | 16.5–18.5 | 11.0–14.0 | 2.25–3.0 | |
| Typical | | | 0.02 | 17.0 | 11.9 | 2.5 | |
| 347 | 347 S17 | W.Nr. 1.4550 | 0.08 max | 17.9–19.0 | 9.0–12.0 | | Nb 10xC/1.00 |
| Typical | | | 0.06 | 18.5 | 10.0 | | 0.75 |
| 321 | 321 S12 | W.Nr. 1.4541 | 0.08 max | 17.0–19.0 | 9.0–12.0 | | Ti 5xC/0.7 |
| Typical | | | 0.05 | 17.6 | 9.4 | | 0.4 |

expansion than carbon steels and coupled with a lower thermal conductivity are prone to unequal expansion and distortion when welded with high heat input arc welding techniques. Nevertheless, partial penetration laser welds in thick section butt joints should be avoided because of possible weld solidification cracking, due to the unwelded portion of the joint resisting the high weld contraction stresses as the weld cools. Also the absence of oxygen or moisture entering the unwelded region of partial penetration butt joints during service, can lead to crevice corrosion which penetrates the weld metal and causes subsequent weld failure.

*Ferritic stainless steel*

These 400 series steels do not possess the good all-round weldability of the austenitic grades, they nevertheless are fusion welded extensively, Table 4.5, and there is no reason why they should not be laser welded. The BSC stainless grades 430, 434 and Hyform 409 have perhaps the best weld performances due to their lower carbon and chromium levels.

Fusion welding of the ferritic grades in some cases impairs joint toughness and corrosion resistance. The reduction in toughness is due in part to the formation of coarse grains in the HAZ and martensite formation which occurs in the higher carbon grades. The low heat input of laser welding, however, reduces the level of grain coarsening in the HAZ when compared with high heat input welding techniques. Nevertheless, the HAZ may have a higher hardness due to the fast cooling rate.

*Martensitic stainless steel*

The martensitic grades produce poorer quality welds then the austenitic and ferritic grades. The high carbon levels, Table 4.6, produce very hard and brittle HAZs and welds, due to the formation of martensite. If carbon contents above 0.1% have to be welded, then an austenitic stainless steel filler material should be used to improve weld toughness and to avoid weld cracking. However, this will not reduce the brittleness in the HAZ. Where practical, cracking and brittleness can be reduced by pre-heating (grades with carbon levels greater than 0.1%) to 350°C and tempering after welding at 650–750°C.

***Table 4.5*** **Ferritic stainless steels (common grades)**

| Specifications | | | | | | | |
|---|---|---|---|---|---|---|---|
| AISI | BS 1449 | Others | C | Cr | Ni | Mo | Others |
| 403 | 403 S17 | W.Nr. 1.4000 | 0.08 max | 12.0–14.0 | 0.50 max | | |
| | | | 0.05 | 12.30 | 0.30 | | |
| 405 | 405 S17 | W.Nr. 1.4002 | 0.08 max | 12.0–14.0 | 0.50 max | | Al. 0.10–0.30 |
| | | | 0.065 | 12.30 | 0.30 | | |
| 409 | 409 S17 | W.Nr. 1.4512 | 0.09 max | 10.5–12.5 | 0.70 max | | Ti 5xC–0.70 |
| Typical | Hyform 409 | | 0.03 | 11.5 | | | Ti 0.3 |
| 430 | 430 S15 | W.Nr. 1.4016 | 0.10 max | 16.0–18.0 | 0.05 max | | |
| Typical | | | 0.055 | 16.5 | | | |
| 434 | 434 S19 | W.Nr. 1.4113 | 0.10 max | 16.0–18.0 | 0.05 max | 0.9–1.30 | |
| Typical | | | 0.05 | 16.9 | | 1.10 | |
| 446 | | | 0.02 max | 23–27 | | | |

**Table 4.6 Martensitic stainless steels (common grades)**

| Specifications | | | | | | | |
|---|---|---|---|---|---|---|---|
| AISI | BS 1449 | Others | C | Cr | Ni | Mo | Others |
| 410 | 410 S21 | W.Nr. 1.4006 | 0.15 max | 11.5–12.3 | 1.0 max | | |
| Typical | | | 0.1 | 12.0 | | | |
| 420 | 420 S45 | W.Nr. 1.4021 | 0.28–0.36 | 12.0–14.0 | 1.0 max | | |
| Typical | | | 0.03 | 12.50 | | | |
| 431 | — | BS Specn. 4.S80 | 0.20 max | 15.0–18.0 | 2.0–3.0 | | |
| Typical | | | 0.15 | 16.5 | 2.25 | | |
| 440 | | | 0.6–0.75 | 16.0–18.0 | | 0.75 max | |
| Typical | | | 0.65 | 17.0 | | 0.50 | |

## Aluminium

Aluminium and its alloys are extremely difficult to laser weld and achieve good weld properties. As these are major engineering materials an explanation for this is presented.

Aluminium is a poor absorber of laser light, it also has a high thermal conductivity and a very short temperature range between liquidus and solidus and when liquid its viscosity is very low. All these factors contribute to poor control of the laser welding action and weld formation. Consequently, laser welding of aluminium and its alloys is difficult and the weld quality is often very poor. Blowholes and porosity are usually the main problems together with weld solidification cracking and a loss of alloying elements, due to evaporation, in some alloys. For full penetration welds, controlling the weld underbead formation is difficult for certain alloys.

High laser power densities are necessary to produce a weld keyhole in aluminium (black coatings to increase beam absorption have been experimented with, but are not practical in production because they can contaminate the weld and are not necessary once the keyhole has formed). Usually, once the keyhole is formed and a short length of weld made, the high thermal conductivity of the material causes a rapid build-up of heat ahead of the keyhole. The hot material ahead of the beam considerably increases the absorption of the laser light, invariably resulting in too much energy being imparted into the weld and causing severe expulsion of molten metal, leaving blowholes in the weld surface.

Another problem which adversely affects welding of aluminium and its alloys is the natural oxide and other contamination on the material surface. Moisture on the oxide and from the atmosphere produce hydrogen which can be taken into the weld and cause porosity. Hydrogen is highly soluble in aluminium and is perhaps responsible for most of the porosity in aluminium welds.

Despite these drawbacks, laser welded seam butt joints which adequately achieve their service requirement are made during the fabrication of the roll formed box sections used in double glazing window spacers. The materials, which are usually one of the 5000 series alloys, in thicknesses of 0.3–0.6 mm, are welded at speeds of up to 100 m/min.[13] For 6000 series alloys, Al-Si filler wire has been shown to produce hermetic welds for microwave packages and small pressure vessels. Nevertheless, in general aluminium and its alloys are unsuitable for laser welding and much more research is necessary before this situation can be reversed. To make quality precision welds in aluminium alloys electron beam welding is considered to be the most suitable power beam technique.

### Copper

In general, $CO_2$ laser welding of copper is not recommended. Copper is an excellent reflector of laser light; highly polished copper mirrors are used in $CO_2$ lasers to transmit the laser light within the laser itself and between the laser and its workstations. Nevertheless, it should be noted that successful micro-spot and seam welds can be made between ($<1$ mm thickness) copper foil and certain other materials when using Nd:YAG lasers employing very short high energy pulses. For quality low distortion welding of copper, especially where deep butt welds are required, electron beam welding is recommended.

## Effects of surface coatings and treatments on the weldability of steel

### Surface coatings

Surface coatings are widely used on steels to provide corrosion resistance and sometimes a good surface appearance. When coatings are to be used on welded fabrications a decision has to be made on whether to apply the coating before or after fabrication. Sometimes it is impractical to apply a coating after fabrication because the product is too large or because the coating material will not enter crevices where, for example, corrosion protection is vital, such as between lap joints in vehicle bodies and in domestic appliance casings. In these circumstances it is often necessary to weld materials with the surface coatings present. Unfortunately, some surface coatings can cause undesirable reactions in the weld metal which reduce weld strength and absorption of the laser beam. In this respect lap joints, where the coating exists across the interfaces to be joined, are more prone to weld formation problems than butt joints.

#### *Electro-plated and hot-dipped (galvanized) zinc*

Laser welding of thin steel (0.8–1.5 mm) metal parts coated with zinc is becoming increasingly popular in the fabrication of agricultural equipment, vehicle body shells, automotive products and domestic appliances. Successful application, however, is dependent on the coating thickness, its uniformity and especially the joint configuration used.

Butt joints, where the zinc coating does not exist at the joint interface, normally present no serious problems to laser welding. As an example, joints, 2 m long, between sheets 0.75 mm thick with zinc coating thicknesses of typically 20 μm have now been reliably produced for several years by Thyssen Stahl for Audi. These welded sheets are subsequently pressed into car floor panels as described in Chapter 3. Much of the zinc coating in the weld region, a strip 1–1.5 mm wide, is vaporised and does not appear to enter and adversely affect the weld metal. Erichsen cupping tests on the welded joint show formability is approximately only 10% below that of the parent metal (the Erichsen cupping test involves forcing a hemispherical plunger into a clamped piece of sheet until the sheet just fractures; the depth of the depression is then measured).

Lap joints between single sided zinc coated steel sheets, where the coatings are on the outside faces, weld similarly to butt joints. However, when the coatings are at the joint interface serious problems which affect weld formation arise. Zinc boils at 900°C and steel melts at nearer 1500°C, consequently, on close fitting lap joints, zinc at the joint interface will have vaporised and started to pressurize before the weld keyhole has formed. Once the keyhole is formed bubbles of vaporised zinc try to escape through it. If the combined coating thicknesses at the joint interface are less than 15 μm and on sheet thicknesses greater than 0.8 mm (to keep the ratio of zinc to steel low) the gas bubbles usually escape before keyhole closure and solidification occurs. Any trapped bubbles will usually be very small and well dispersed along the length of the weld.

Zinc coating thicknesses greater than 15 μm between the faces of close fitting lap joints produce a sufficiently large volume of zinc vapour that it cannot all escape before keyhole closure occurs. In fact the pressure build up within the gas bubbles causes them to blow out the molten metal before it can solidify. Figure 4.3 shows sections of welded lap joints between steel sheets with double sided zinc coatings of 8 and 12 μm thick and the difference in weld formation is evident. Slowing down the welding speed does not solve the problem of zinc vapour outgassing, because more heat is conducted into the work and this vaporises more zinc from a wider area. Improvements can be achieved by pulsing the laser beam. In this way, the short duration pulses (e.g. 0.5 ms at 200 Hz) considerably limit the interface surface

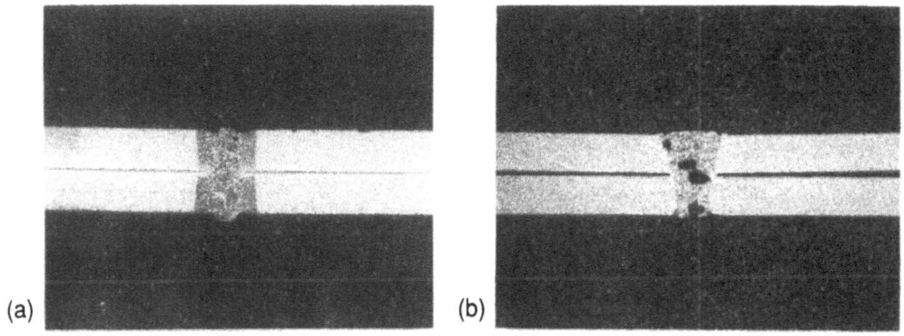

**4.3** Cross sections through $CO_2$ laser welded lap seam joints between 0.7 mm thick steel sheets with double sided zinc coatings of (a) 8 μm and (b) 12 μm. This illustrates the effect of a small increase in zinc coating thickness on weld formation.

area over which zinc is vaporised and reduce the volume for ejection through the keyhole. The new generation of high power (>1 kW) Nd:YAG lasers looks promising in this respect, as they can easily deliver the necessary short high energy pulses.

Some success in welding zinc coatings of greater than 15 μm at the joint interface has been achieved by maintaining a small gap, typically 0.1 mm, between the lap joint faces, thus allowing the zinc vapour to exit between the sheets as well as through the weld keyhole. Under laboratory conditions, where care can be taken to control the gap size, this approach has proved relatively successful using a clamping method described in Chapter 6 (Fig. 6.4) however, under production conditions involving large pressed steel parts where the lap joint may follow a curved surface, controlling the gap size is extremely difficult. Nevertheless, this approach has been patented.[14]

When welding lap joints with zinc coatings at the interface and with the laser operating in the continuous wave mode, the zinc vapour ejected from the keyhole reacts with the laser beam and forms a dense plasma cloud above the workpiece. The plasma, which is partially opaque to the laser beam, reradiates the laser beam over a wider area than the original focused spot size, thus reducing the power density at the work and also melting the zinc coating over a wider area each side of the joint line. Consequently, weld penetration is reduced and the increased heat transfer into the surrounding material is sufficient to cause sheet buckling if the welding speed is too slow. For example, when making lap seam joints between 0.8 mm steel sheets, coated with $<10$ μm of zinc, welds made at 2 m/min which penetrate both sheets can cause noticeable buckling. To eliminate the buckling it is necessary to increase the welding speed to at least 3 m/min (5 m/min is the norm). Moreover, in general higher welding speeds and powers (15–20% power) than for bare steel, together with the smallest practical focused beam spot size are necessary to achieve acceptable results when welding lap joints with zinc coatings at their interface.

A most important factor when laser welding zinc coated steels in production is the uniformity of the coating thickness. When specifying a coating thickness it should not be an average, but as close as possible to that required and not exceeding it. In this way, more confidence in weld quality will be achieved. New batches of material should always be carefully checked for coating thickness. It is not unusual for suppliers when out of stock of one thickness to supply the next thickness up without any extra charge, believing they are doing the user a favour. Electro-plated zinc coatings should be examined carefully because although the surface may look uniform to the naked eye, the plating thickness can vary considerably. 'As dipped' hot-dip galvanize is not usually suitable for lap joints as the coating is usually too thick, however, that which is post-dip processed to control the thickness has a uniform thickness and is generally quite suitable providing the coating thickness is correct.

Finally, zinc vapour can be damaging to the laser focusing optics and therefore an air knife, described in Chapter 11, placed near the focusing optic to keep vapour away is recommended. General fume extraction in the weld area is also necessary to protect the operator, see Chapter 12.

*Sherardised (zinc coating)*

Despite the fact that much thinner surface coatings can be obtained with sherardising than with hot-dip galvanizing and a more even film is obtained, the weldability of sherardised steel is poor. This is because sherardising is a cementation process, similar in many respects to carburising, in that the zinc is made to combine with the steel surface by heating the work in zinc dust. This causes the zinc not only to coat, but also to diffuse into the surface of the steel. When welded the zinc boils and outgasses and seriously disrupts weld formation. Since sherardising is usually used on small components the possibility of welding before sherardising should always be considered.

*Nickel plate*

Nickel plated steels are not widely used for welded fabrications. For laser welding the most common application is perhaps electronic packages.[15] In this instance, the packages are usually nickel plated iron-nickel-cobalt alloys, such as Kovar. Electrolytic and electroless plating techniques are used. Welds on electrolytic nickel plated Kovar are prone to very fine surface cracks which are usually very shallow, running transverse to the direction of welding and which are thought not to be too detrimental to weld properties. Electroless nickel platings, however, produce serious solidification cracking when welding in the continuous wave mode. These cracks can extend through the full thickness of the weld and propagate into the HAZ. They can be reduced by pulsing the laser beam and satisfactory results have been achieved using an Nd:YAG laser pulsed at 150 Hz and using 1 msec pulse lengths. The difference in weldability of the two types of nickel coating is attributed to the presence of phosphorus in the electroless nickel.

*Aluminised*

Aluminised steel sheet (aluminium + small % silicon) is popular for the fabrication of domestic appliances. It can be laser welded and the surface appearance and weld properties, in both butt and lap joints, are satisfactory for most applications. A mass produced, laser welded, gas cooker top burner manifold fabricated from thin aluminised steel sheet is described in Chapter 3.

In thin sheet steel lap joint configurations the laser weld formation is not affected by outgassing of the coating material. This is because aluminium boils at 2450°C which is approximately 1000°C higher than the melting point of steel. The resultant weld, Fig. 4.4, has large grains and usually a central and almost linear continuation of grain boundaries running through the thickness of the weld. This, together with the large grains, weakens the weld in relation to the parent metal strength; a tensile-shear test on the specimen shown in Fig. 4.4 produced a failure at the edge of the weld between the two sheets and the strength was 15% lower than the parent material. Nevertheless, the strength is more than adequate for many applications.

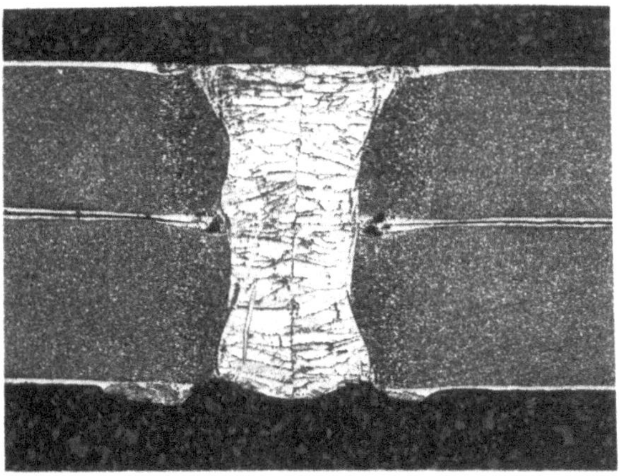

4.4 A section through a $CO_2$ laser welded lap joint between two 0.8 mm thick steel sheets, each having a double sided aluminised coating of 15 μm. Welding power 5 kW, speed 6 m/min.

*Cadmium*

Laser welding or any other method of fusion welding cadmium plated steel is not recommended. This is because cadmium boils at 760°C and produces a poisonous vapour. Also cadmium films which have poor strength may form around the weld grain boundaries and produce cracking.

*Gold*

Gold platings are rarely used on steel but are common on the iron-nickel-cobalt alloy microcircuit packages used in the electronics industry. These materials laser weld extremely well if a thin gold flash is used, but thick platings can cause weld solidification cracking.

*Paint*

Painted surfaces are not easy to comment on as there are many different types of paint and applied thicknesses which can react differently when heated and burnt.

The grey autobody primer paints when placed between lap joints promote porosity and blowholes in the welds. Butt joints, however, where only the sheet surfaces are painted, have been welded satisfactorily.

Epoxy based primer paints, 20 μm thick, at the joint interface of T butt joints between 6 mm thick steel plates have been reliably laser welded under experimental conditions.[16]

Generally, when painted coatings are present, the quality of the weld is related to the vapour produced, the ratio of the coating thickness to the sheet or plate thickness and the joint configuration. Unfortunately, under these circumstances one is usually left to conduct one's own experiments to determine the success.

## Surface treatments

Surface treatments are widely used on steel to harden the surface and provide good wear resistance whilst retaining toughness of the core material. Gears and bearing surfaces are common recipients of surface treatments.

Laser welding is widely used for fabricating gear clusters and other vehicle drive train components, such as clutch cups and differential drive housings. During the design stage the question often arises as to the weldability of components which have undergone surface treatments. Carburising and nitriding are the most common methods of hardening these components and both present serious problems to fusion welding techniques. The ideal approach is to conduct these surface treatments after welding, which unfortunately is not always possible. The problems associated with the welding of these surface treatments and the necessary actions to overcome them are discussed below.

### *Carburised*

The cementation process which produces a carburised surface diffuses carbon well into the surface of the component, so much so that the carbon content at this position is several times higher than the untreated material. For example, a 0.2 carbon steel can, after treatment, have a surface carbon content of 0.6%. Such high levels cause weld solidification cracking and shrinkage cracking in the HAZ, see Fig. 4.5. The only solution to this problem, if the component cannot be surface treated after welding, is carefully to mask off the parts to be joined for a distance of at least 8 mm each side of the joint to prevent the carburising action creeping under the mask and occurring in the weld and HAZ regions.

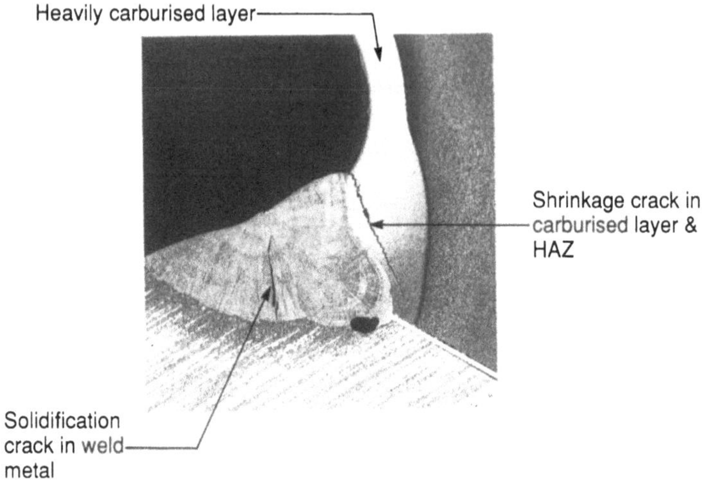


**4.5** Weld solidification crack and HAZ crack caused by the high carbon content present in a deep carburised layer on one of the component parts. The weld shown is a conduction limited laser weld.

*Nitrided*

Nitrided surfaces are generally unsuitable for fusion welding processes. A weld made over a nitrided surface will remove the surface hardness in the vicinity of the weld. Moreover, the resultant weld will be porous near the surface and may also have cracks.

## Possible weld defects and solutions

A good weld is a weld which adequately fulfils all its service requirements. Therefore at the outset of choosing welding for joining parts together it is essential to establish what the weld service requirements will be and whether the chosen welding technique will produce a weld which will achieve them; all easier said than done, but essential for the production of a reliable product.

A low stressed weld on a light duty piece of garden furniture can often tolerate numerous defects and a deliberately made larger weld than is necessary will often fulfil the service requirements and life expectation. On the other hand, a weld used in a highly stressed situation on a vehicle frame, drive train or engine part will require careful calculation for size and then subsequent mechanical testing and metallurgical examination to ensure the service requirements will be met. However, this is not the end of the story because weld defects can occur due to unforeseen material changes and production errors. Most potential defects of this nature can be avoided by taking steps to make sure they do not occur in the first place. To do this an awareness of the common weld defects is necessary and therefore these will be discussed, together with basic solutions where possible.

The defects to be described should not be viewed as a blacklist for not using welding. Welding correctly applied is remarkably reliable, after all every day of our lives, often without realising it, we rely on and are in contact with welded items some of which contain hundreds of welds and they rarely let us down.

### Porosity

Porosity occurs when bubbles of gas are trapped during weld solidification. Examples of this were described earlier in this chapter in relation to oxygen levels in steel and when making lap joints in certain coated steels. There are, however, numerous other sources which can produce porosity such as the intake of air and moisture into the weld pool. These are easily dealt with by effective gas shielding, which is described in Chapter 5. Another source of porosity is surface contamination on or about the joint interface, common contaminates are oil, grease, moisture and paint. All of these can be eliminated by careful cleaning procedures, also described in Chapter 5. Fine porosity (pores less than 0.2 mm diameter) is present in most fusion welds and normally in the form of almost spherical cavities which apply insignificant stress to the weld. The most serious problem they pose is loss of weld cross section which can occur when they are in large numbers and in close proximity. Pores which are very small in relation to weld depth and which are well dispersed over the weld length will not usually significantly affect the mechanical properties of the weld. For example, a welded butt joint between 1 mm thick sheets with a 0.2 mm diameter pore dispersed on average every 10 mm along the weld length, will suffer a loss of longitudinal cross section of less than 0.5%. Nevertheless, from this example *do not* assume that pores of a diameter equal to 20% the sheet thickness are always acceptable. Very large pores representing this percentage in thick sections could contain undesirable levels of gas, such as hydrogen, which could cause component in-service failures at a later date.

Deep laser welds, in close fitting butt joints, which do not fully penetrate the through thickness of the joint (blind welds) are prone to porosity, Fig. 4.6. This is due to material outgassing and the bubbles having too long an escape route from the root of the weld to the surface, before keyhole closure occurs. Where a keyhole penetrates the full through thickness of a joint, any gas bubbles are provided with two exits and a shorter escape route. Nevertheless, the combination of welding power and speed is important to ensure the escape of gas bubbles.

### Solidification cracking

Solidification cracking, Fig. 4.7, which is also referred to as hot cracking, occurs at high temperatures above the solidus under circumstances where the material has low ductility and is under high contraction stresses.

Solidification of the liquid weld metal begins at one temperature (the liquidus) and completes at a somewhat lower temperature (the solidus). Between the two levels the alloy forms an incompletely solidified mass which is brittle and has little or no ductility until cooled to lower temperatures. The brittleness is present in the liquid films which surround the solidifying grains or dendrites and when subjected to the high transverse contraction stresses, during weld cooling, solidification cracking may result in the film boundaries; usually those near the centreline of the weld which are the last to cool. When high carbon levels are present, weld brittleness occurs, due to the carbon forming martensite which is a brittle and weak substance in the final cooling stages of the weld.

In steel, the main element which contributes to hot cracking is sulphur which promotes the formation of low melting point iron sulphide films. Other elements are phosphorus and boron. The sulphur in steel can be rendered harmless by additions of manganese which promote the formation of manganese sulphide. Likewise, high levels of carbon can also be rendered harmless by the addition of manganese.

In short, the risk of solidification cracking in carbon steel weldments can be minimised by:

- Maintaining a low carbon level in the weld zone – if necessary by the addition of a low carbon-manganese material;
- Keeping sulphur and phosphorus levels as low as possible; ideally a combined level below 0.02%. If this is not possible the addition of a low carbon-manganese material to the weld should be used;

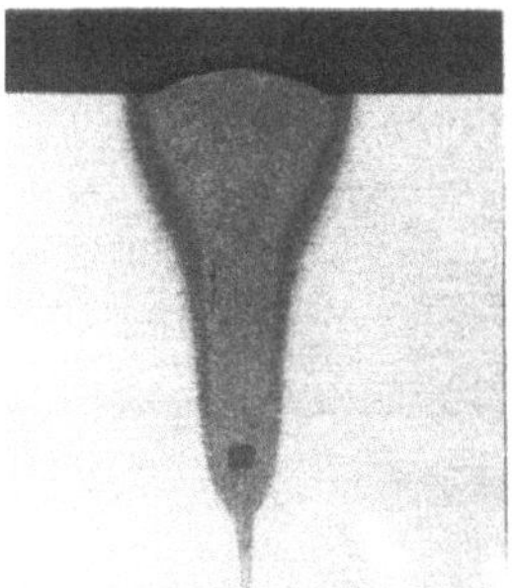

**4.6** An example of root porosity, which is not uncommon in deep and narrow blind welds made with power beams (electron beam and laser welding). The weld depth is 15 mm.

**4.7** A weld cross section showing a typical solidification crack towards the centre of the weld, which has helped promote a cold crack to run from the notch left by the unwelded region of the joint interface. In this case the cracks were caused by the high sulphur level in the steel and the joint design producing a stress raising notch at the weld root.

- If the materials are prone to cracking and the addition of suitable elements to modify the weld is not possible, partial penetration welds should be avoided and, where possible, modifications to the joint design should be made to reduce the weld contraction stresses on cooling. Also, to combat contraction stresses if possible, the weld faces should be held together under pressure during the weld cycle and in thick sections pre-weld tack welds should be made along the joint length, see Chapter 7.

### HAZ liquation cracking

The low melting point grain boundary films described above may also be present in the HAZ and may melt and form fine cracks under the influence of the thermal stresses induced during welding. Laser welds are not normally susceptible to this form of cracking in weldable grades of steel because of the low heat input.

### Cold cracking

The most common form of cold cracking is what is known as 'hydrogen induced cold cracking' and this can occur in ferritic, martensitic and high alloy steels. The cracks usually form in the HAZ (see Fig. 4.5) after welding but can occur in the weld metal itself. The cause is usually a combination of three factors:

- The presence of hydrogen;
- A brittle martensitic structure;
- The presence of high contraction and residual stresses.

The problem may be avoided by following the possible remedies for solidification cracking. Where this is not possible pre-heating to reduce the cooling rate and tempering as described earlier in this chapter for martensitic stainless steel may be applied.

### Stress corrosion cracking

Stress corrosion cracking is a failure mechanism which occurs through the combined actions of corrosion and tensile stress. Partial penetration welds which have root crevices are prone to this type of failure and should not be used in corrosive environments, such as hot nitrate solutions and ammonia storage tanks. In such environments, carbon and stainless steels are particularly susceptible. The type of cracking can be transgranular and intergranular and is dependent on the corrosive environment. Such failures require expert interpretation.

In closing this section it should be noted that defects discovered in welds where high mechanical properties are required should never be ignored. If the discoverer is not an expert on the subject, advice should always be sought from professional consultants at a welding or materials joining institute or a university where knowledge of welding metallurgy and weld properties exists.

## References


1 As Ref. 1, Chapter 2.
2 Breinan E M and Banas C M: 'Preliminary evaluation of laser welding of X-80 arctic pipeline steel'. Res Council, 1974, No. 201 pp 45–47.
3 Breinan E M and Banas C M: 'Fusion zone purification during welding with high power $CO_2$ lasers'. UARL Report No. R111087-s, United Aircraft Corporation, Connecticut, April 1975.
4 Breinan E M and Banas C M: 'Fusion zone purification during welding with high power $CO_2$ lasers'. In Advanced Weld Tech, 2nd Intl Symp. of JWS, 25–27 August 1975, Osaka.
5 Parrini C, Banas C M and deVito A: 'Laser welding of pipeline steels'. In 'Welding of HSLA (microalloyed) structural steels', Proc Intl Conf, 9–12 November 1976, ASM Rome.
6 Stoop J and Metzbower E A: 'A metallurgical characterisation of HY-130 steel welds'. *Weld J*, 1978, Vol 57, No11, pp 345–353.
7 Willgoss R A, Megaw J H P C and Clark J N: 'Assessing the laser for power plant welding'. *WMF* 1979, 47 (2 March), pp 117–127.
8 Metzbower E A and Moon D W: 'Mechanical properties, fracture toughness and microstructures of laser welds in high strength alloys'. In 'Applications of lasers in materials processing', Proc Conf, Washington, 18–20 April 1979.
9 Metzbower E A and Moon D W: 'Properties and structures of laser welds of high strength alloys'. In Weldments: 'Physical metallurgy and failure phenomenon', Proc 5th Bolton Landing Conf, August 1978, General Electric Co, Schenectady, USA
10 Breinan E M, Banas C M and Greenfield M A: 'Laser welding – The present state of the art'. IIW Doc IV-181-75.
11 Watson M N: 'Laser welding of a 12.5mm thick niobium microalloyed structural steel at powers of up to 9 kW' Weld Inst Res Report 243/1984.
12 Duhamel R F and Banas C M: 'Laser welding of steels and nickel alloy'. Second Int Conf on 'Applications of lasers in materials processing'. ASM Conf Proc, 24–26 January 1983.
13 Thorstensen B: 'Laser welding of aluminium'. The Industrial Laser Annual Handbook, 1989 Edition, Pennwell Publishing Co. ISBN 0-87814-341-6.
14 'Joining distance between two coated sheets'. European Patent Application 337 182 AZ. Filed 28 March 1989. Pub 1:18 Oct 1989. 4 fig, 4 claims.
15 Stockham N R and Dawes C J: 'Resistance seam and laser welding of large hybrid metal packages'. *International Journal for Hybrid Microelectronics*, Volume 6 Number 1, October 1983.
16 Martyr D R: 'The application of high power laser technology to ship production'. North East Coast Institution of Engineers and Shipbuilders, 12 Windsor Terrace, Jesmond, Newcastle Upon Tyne NE2 4HE.

Chapter 5

# Laser welding parameters and their effects

Weld shape and depth are determined by the manner in which the welding energy is applied to the joint. For laser welding, the energy input is controlled by the combination of the following parameters: focused spot size, keyhole shielding gas, laser power (power is combined with pulse width and frequency in the case of pulsed lasers) and welding speed. These being correct, the repetition of welding performance then depends on the material preparation, joint fit-up and laser beam to joint alignment.

All the above parameters are covered in this chapter, together with the important related aspects of: focus (f) number; depth of focus; focus position; shielding gases; gas shielding devices; when and how to use a plasma suppression device; how to establish welding conditions and tolerances; and finally a troubleshooting check list.

## Focused spot size, depth of focus and focus position

### Focused spot size

To achieve the required power density for keyhole welding ($10^3$ to $10^5$ W/mm$^2$) selection and maintenance of the focused spot size is most important. Consequently, care is necessary in selecting the characteristics of the focusing optic which determine the focus spot size.

The simple lens will be used for the purpose of explaining the focusing optic characteristics which have to be considered when establishing the focus spot size. However, there are several different optical equipments available for focusing laser beams and these are discussed in Chapter 11.

When light is focused, the rays converge to a very small waist diameter, d, and length, l, Fig. 5.1, before diverging again. The exact minimum waist diameter and length achieved depends on the type of optic; its

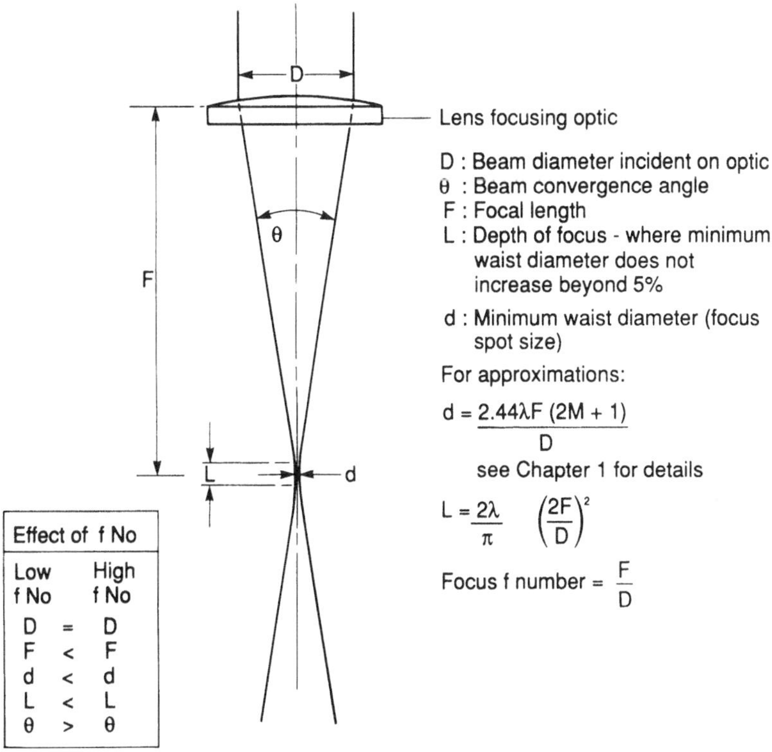


5.1 Characteristics of a focused laser beam.

focal length, F; the beam diameter, D, incident on the optic; whether the incident beam is converging or diverging; the beam TEM number; the light wave length and the laser power.

The actual minimum waist diameter, d, of focused laser light produced is almost impossible to calculate accurately or physically measure, even when using the special scientific equipment described in Chapter 9. The spot diameter calculation described in Chapter 1 only gives an approximation since it does not take into account all the controlling factors mentioned above, some of which in themselves are difficult to establish and can vary as the laser power is increased. Furthermore, the formula on its own will not establish the focus spot size which should be used. It may at first appear that one can easily establish the spot diameter for a given laser power, because the required power density range is known, and then establish the required optic focal length from the spot size formula. This approach is more difficult than it sounds because the power density range estimated for keyhole welding is wide and the focus spot size in isolation does not contribute to good welding performance.

Once welding power, speed and gas shielding requirements have been met, the focused convergence angle, θ, depth of focus, L, and the focus spot size, d, collectively control welding performance. As they are interrelated they cannot be individually adjusted and a compromise has to be established. The main controlling factors of the relationship which are in the hands of the operator, as far as the focusing optic is concerned, are the incident beam diameter and the focal length. The dimensional relationship and effect of these two factors is conveniently expressed as the f number, which is derived by dividing the focal length, F, by the incident beam diameter, D.

From Fig. 5.1, it will be seen that the lower the f number the smaller the beam waist diameter, d, and depth of focus, L, and the larger the convergency angle, θ. Although low f numbers produce smaller spot sizes and therefore higher power densities which enable higher welding speeds than with higher f numbers, they do not ensure such good all-round welding performance. This is because the welds are narrower thus making joint fit-up and beam on joint alignment more critical; more weld metal can be lost through evaporation producing undercut welds; the shorter length, or depth of focus, makes the setting of the focus position more critical and less tolerant to variation; the larger beam convergence angle reduces the laser beam's ability to access certain joints and narrow gaps; and finally the shorter focal length, F, requires the optic to be closer to the work, which makes it more prone to weld spatter, thermal and vapour damage.

The experience of many $CO_2$ laser users and equipment suppliers has shown that f numbers from 6 to 9 generally provide the best all-round welding performance for powers up to at least 10 kW. To achieve very high speed welds it is possible to use numbers down to f3, but due to reasons described in the previous paragraph, precision process control becomes necessary for reliable welding performance. Below f3, beam focus ability deteriorates especially when using a lens, because of spherical aberration which will change the waist position and increase its diameter. Numbers greater than f10 generally produce too large spot sizes and need very high laser powers to achieve the required power density and thus make inefficient use of the laser power.

## Selecting the focus spot size

Unless welding speed is paramount, the beginner is advised to choose the focused spot size for welding based on f number 7.5 for $CO_2$ laser welding (f4 is a practical starting point for Nd:YAG). All that is necessary is to measure the incident beam diameter at the proposed working power, at the position of the focusing optic, then multiply the diameter by the f number to establish the focal length of the optic required. (The incident beam diameter can be established for Nd:YAG lasers by taking a photographic print and for $CO_2$ lasers by an acrylic burn print, both methods are described in Chapter 11.) If the focal length dimension does not coincide with a manufacturer's stock size it should be rounded up or down to the nearest dimension; rounded down if welding speed is important. Armed with beam diameter, the required focal length and beam power, the suitable focusing equipment (Chapter 11) for the welding application can be chosen.

## Depth of focus

From Fig. 5.1 it can be seen that the focused laser beam has a waist length and minimum diameter which increases with f number. The waist length is very useful for laser welding, since the proportion of it where the diameter has not increased beyond 5% of the minimum waist diameter, provides a working depth of focus in which the power density will decrease no more than 10% and which can be easily accommodated by laser power selection. Thus, it provides a tolerance for the setting of the focus position on the work and thus enhances the reliability of weld formation.

There are numerous formulae for approximating the working depth of focus of a focusing device and one is given in Fig. 5.1. All such formulae are influenced by the factors which control the focus spot, which were discussed earlier. They do, however, provide a dimensional indication of the variation of focus position (minimum waist) about the work surface which should be experimented with to establish the real working range, i.e. where weld depth is acceptable, and hence the actual practical tolerance to variation in focus position.

To establish the working depth of focus practically the operation is simplified if one can first ascertain the approximate minimum waist diameter position, since this gives a starting position (dimension) each side of which should be investigated. In theory this should be at a distance from the focusing optic which is equal to the focal length, however, the manufacturing tolerance is often only to within ±5% depending on the optic supplier. Furthermore, for a transmissive optic (lens) the true focal length starts from a point within the optic (Chapter 11) and not at a point on a surface face (some optic suppliers provide a dimension from the front or back face of the optic for a given focal length) and unless the optic has been precision manufactured for a specific laser gun, where a datum point can be taken to the focal position using an established dimension, the minimum waist diameter needs to be found. Even where a measurement from a datum point exists it is an advantage to check it practically in case the optic thermally expands at the laser working power and causes thermal lensing, which will change the minimum waist position and focus spot size.

To find the minimum waist position for an Nd:YAG laser a small steel plate should be positioned with its surface set at the theoretical lens focus distance and a single pulse spot made on the plate surface. The plate should then be moved along and other single pulse spots made with the plate surface raised in 0.1 mm steps each time. After making four or five spots, their diameters should be measured under a microscope. If the spots gradually get smaller and then increase, the smallest spot will represent the approximate minimum waist position and its surface position setting should be returned to, and with a vernier height gauge a measurement should be taken from the plate surface to a practical datum point on the laser gun and recorded. Should the spots continue to get larger then the exercise should be repeated with the plate surface lowered from the original starting position. If possible when making the single spots, the approximate average pulse power and pulse length to be used for welding should be employed because the beam focus position can alter as the power is changed. Also a plate thickness slightly more than the expected penetration depth should be used to avoid damage to the work table. An alternative method of finding spot size is to make spot burn holes in polyimide films and measure their diameter.

For $CO_2$ lasers with powers up to around 6 kW, a satisfactory way of establishing the minimum waist position is by making a focused beam print on an acrylic sheet as shown schematically in Fig. 5.2. It is important when using this method to remember to take the print at the laser power to be used in case the focus position moves with changes in power. It is also advisable to traverse the acrylic sheet through the beam at 4–5 m/min to avoid unwanted burn print distortion through heat conduction which would occur at lower speeds.

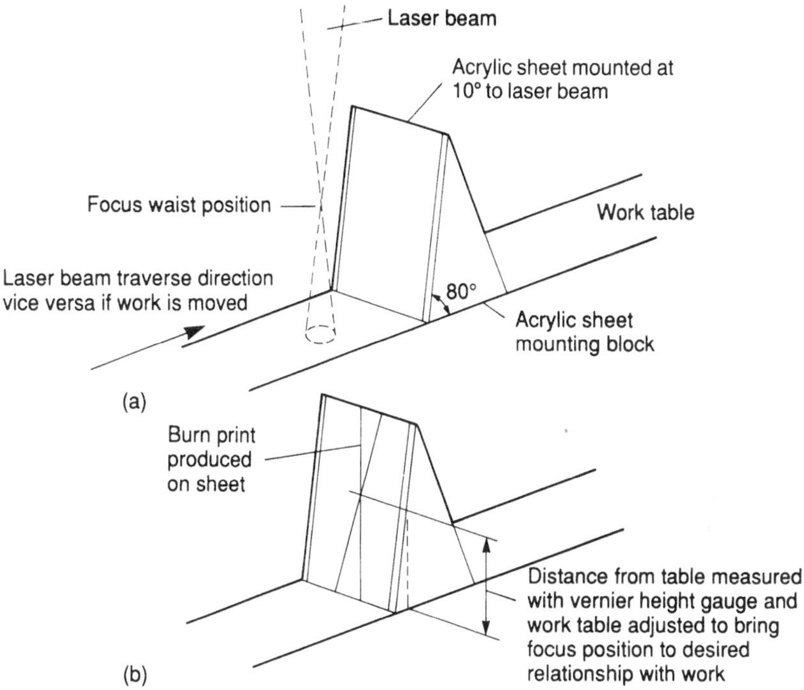


**5.2** Principle for establishing the minimum focus waist position by using an angled acrylic sheet: a) Before beam print; b) After beam print.

Once the minimum waist position is established and ideally in relation to a fixed datum point on the laser gun for future reference, laser melt runs can be made to find the practical working depth of focus based on the penetration depth achieved. (The term melt run is used when a welding condition is used on a single plate, i.e. without providing a joint line of abutting faces.) The melt runs are perhaps best conducted with the aid of low carbon steel coupons each about 50 mm long × 30 mm wide and a few millimetres thicker than the expected weld penetration depth for the welding power and speed to be used in production. The runs should be made along the 50 mm length and down the centre of the 30 mm width. When making the melt runs, five coupons are required: one should be positioned with its upper surface set at the laser beam minimum waist position; two coupons should be positioned to provide surface height positions at slightly more than the extremities of the calculated working depth of focus (Fig. 5.1); and the remaining two coupons at surface height positions halfway between the minimum waist and the extremities, see Fig. 5.3. Ideally, the five

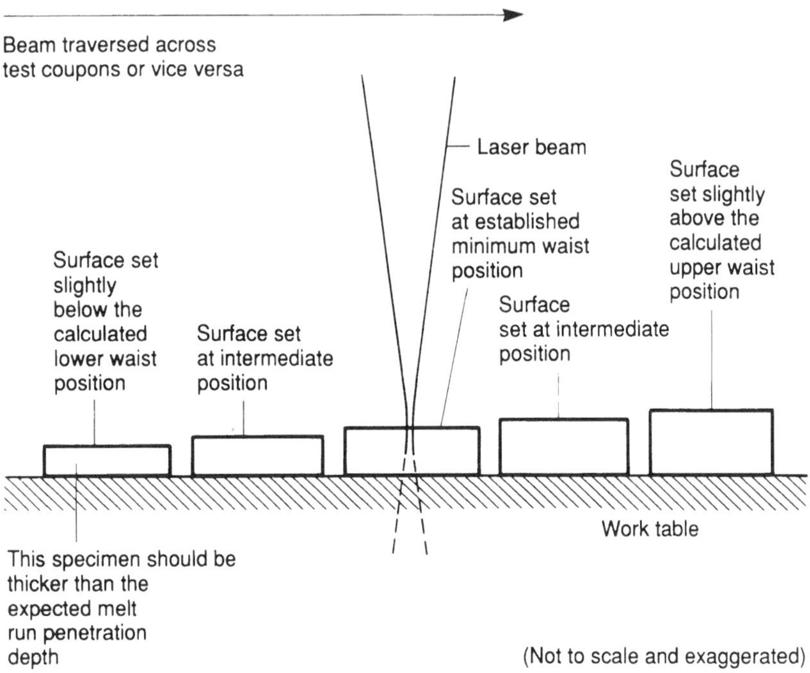


**5.3** A method, employing different thickness test coupons, for finding the practical working depth of focus and possible working tolerance.

coupons' thicknesses should be machined to provide the variations in surface height. The advantage of this is that no confusion over the coupons' surface position occurs and by lining the coupons end to end all the melt runs can be made in one operation (for this Nd:YAG lasers should be operated in a multi-overlapping spot mode). After making the melt runs the coupons should be sawn in half, transverse to the direction of welding, to expose the face containing the melt run penetration depth. The face should then be polished and etched (a simple procedure is described shortly) to expose the melt depth which can then be measured with vernier calipers; a magnifying glass or microscope may be necessary to view the depth of welds made at low laser powers. Once measured the melt run depths can be plotted against their coupon surface height settings to show the relationship. From this the variation in the working depth of focus which will maintain the required weld depth can be established.

On workstations which cater for rotating components where the laser beam is at 90° to the axis of rotation, rectangular test coupons cannot be used. Under these circumstances, stepped round bar test specimen should be used.

Finding the minimum waist position at $CO_2$ laser powers much above 6 kW is difficult by the acrylic burn print method because of excessive melting causing poor waist edge definition and single spots on a plate are misleading because control of the necessary short pulse is difficult. Therefore, for higher powers, the working depth of focus test coupon which would normally have its surface set at the minimum waist position found by an acrylic burn print, should be set to the theoretical focus position, i.e. the focal length of the optic. With this approach it may be necessary to make a second series of tests, or initially use more than five coupons for the melt runs, because the maximum depth of penetration, indicating the minimum waist diameter will probably occur in a coupon other than that which has its surface set to coincide with the theoretical focus position. The alternative is to use a long shallow tapering wedge shaped coupon, however, this approach presents practical difficulties in taking and polishing sections close to and in relation to the calculated working depth of focus extremities.

Low carbon steel is advised for test coupons because it can be easily sawn to expose the melt run penetration depth and if grinding and polishing facilities are not readily available it can be easily filed or machined and then polished by rubbing on a piece of wet silicon carbide paper, laid on a piece of flat glass, and working through the paper grades (grades 120, 320, 400 and 600) if necessary. Moreover, to expose the melt run depth more clearly steel can be easily etched by dipping in a solution of Nital (2 ml nitric acid; 98 ml industrial methylated spirit) and then rinsing thoroughly in water and drying. During the etching operation, eye protection and rubber gloves should be worn.

## The focus position

The focus spot, or the minimum waist diameter position about the work surface, has to be carefully selected to ensure the correct laser power density to form the weld keyhole. Among some researchers there appears to be some conjecture, when making autogenous welds (i.e. those made without the addition of a filler material) whether the position should be on the surface or slightly below it. Whichever the case may be, straying outside the working depth of focus in relation to the workpiece surface is unwise, because small distances will cause large variations in beam diameter, especially with low f numbers where the beam convergence and divergence angles are large. The problem is illustrated schematically in Fig. 5.4 and the point to bear in mind is that small changes in diameter produce large changes in area which produce a corresponding decrease in power density and the potential loss of keyholing ability. For example, if the focus spot size were doubled for a given laser power the area would be four times as large and the power density would be only a quarter of that achieved with the original spot size.

Where should the focus position be? It should be the position which gives the maximum penetration depth and which can be taken from the results obtained when establishing the working depth of focus. From experience this usually appears to be at the minimum waist diameter or just to one side of it, which suggests that either the minimum waist position was not established exactly in the first instance or indeed the best focus position is to one side of it. Whichever the case may be, the position should always be established practically to ensure the best process tolerances and therefore whether the true minimum

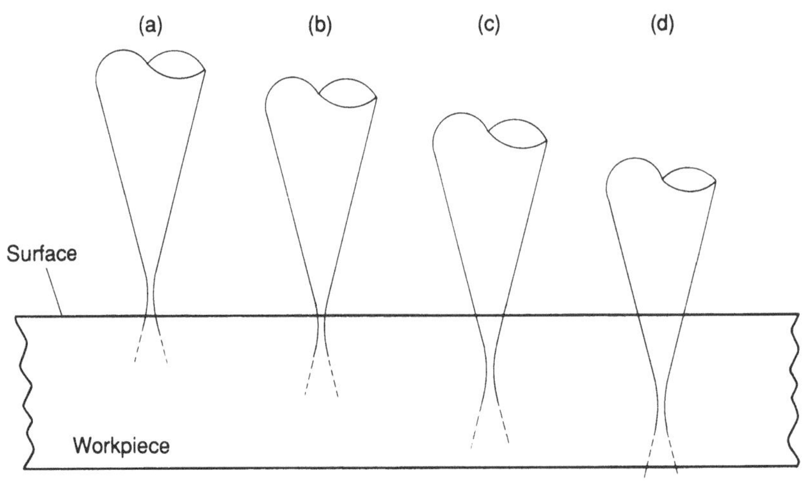


**5.4** The effect of the focus position of the laser beam on the work surface when straying outside the working depth of focus. The incident beam (a) is set on the work surface at the lower extremity of the working depth of focus. Beam (b) is at the upper extremity and the beam diameters (a) and (b) on the work surface are similar. Beam (c) has the upper extremity moved down into the work a distance equal to the working depth of focus and (d) to twice the distance. In both cases the beam diameter incident on the work surface is increased considerably in relation to (a) and (b). These small movements would cause dramatic changes in laser power density and if they were usable they would be extremely difficult to reset.

waist position is on the work surface or not is perhaps really of no major significance.

When making welds which require the addition of a filler wire the working position of the focused laser beam waist has to be considered in relation to the filler wire feed position, joint gap and depth. Because a joint gap is necessary when using filler wire, keyholing is unnecessary, and in its place conduction heating is required to melt the joint side walls and assist feeding of the filler wire into them. Consequently, depending on the joint geometry, radiation of a focused or defocused laser beam via a plasma cloud becomes important. Since the selection of focus spot size and position have to be considered in relation to several other variables when welding using filler wire, they are discussed in Chapter 8.

## Shielding gases, shielding devices and plasma suppression

### The role of the shielding gas and selection of gases

In laser welding the shielding gas has two roles:

- To protect the weld keyhole and solidifying molten metal from oxidation and thus avoid porosity and oxide inclusions which give rise to poor weld quality;
- To protect the transmission of the laser beam as it comes to focus on the work and thus ensure good weld penetration by minimising beam expansion and scattering, which can be caused by vapours and gases around the weld keyhole.

The first of these roles is achieved by several gases, but the second is more difficult. During welding metal vapour ejected from the keyhole absorbs laser power, ionises and forms a plasma cloud just above the weld keyhole. The cloud forms as a brilliant blue/white ball a few millimetres in diameter and at high laser powers it is extremely difficult to look at without viewing through a tinted glass. It is also only partially transparent to the laser beam and if not considerably reduced will expand and scatter the laser beam from its intended path, causing a reduction in weld penetration depth due to the resulting reduction in power density. Under the right conditions the correct shielding gases suppress plasma formation and ensure maximum transmission of the laser beam and its absorption by the workpiece.

The common shielding gases used for laser welding are argon, carbon dioxide, helium and nitrogen (oxygen-free nitrogen referred to as OFN). These different shielding gases do not have the same effect when used with $CO_2$ and Nd:YAG lasers. By comparison the role of plasma suppression is negligible for Nd:YAG lasers because of the different light wave length and the pulsed mode of operation, which appears to limit plasma formation. The shielding gases and devices for these two types of laser will therefore be described separately.

*Shielding gases for Nd:YAG laser welding*

Quite often, satisfactory single spot welds can be made with Nd:YAG lasers with no shielding gas at all. This is because the weld is molten for such a short time that no detrimental oxides can enter it before solidification occurs. However, the surface of the spot does oxidise and soot from metal vapour will also be present. When making continuous seam or butt welds with overlapping spots any oxide and soot contamination on the previous spot will enter the weld pool of the next and therefore for such welds a gas shield is necessary to prevent such contamination and maintain weld quality. For this purpose argon or nitrogen (nitrogen can cause embrittlement in certain steels) is generally used with lasers rated up to about 300 W. Above this power level gas shielding becomes more critical and can affect weld penetration depth and appearance. However, users of high power Nd:YAG lasers operating in the 1 kW average power range have reduced the problem of penetration by using a 20% carbon dioxide, 80% argon gas mixture or argon plus 1–2% oxygen mixture and these are suggested for power levels above 300 W. These gas mixtures produce slight oxidation of the weld surface, however.

Nd:YAG laser welding trials in Japan,[1] with helium and nitrogen shielding gas, have shown similar weld penetration depths are achieved by both gases, but more weld porosity occurs with helium.

*Shielding gases for $CO_2$ laser welding*

When welding with $CO_2$ lasers the lack of weld penetration caused through plasma formation is only prevalent when using the wrong shielding gas or when using combinations of high laser power and low welding speed; usually less than 1 m/min and definitely at 0.7 m/min and below. In the case of too low welding speed a compound gas shield is used, comprising a main shield and a special plasma suppression jet, which is described later in this chapter. Where the welding speed is above 1 m/min efficient plasma control is achieved with simple shielding devices and the selection of the correct types of gas. Accordingly, since most $CO_2$ laser welding occurs above 1 m/min, gas selection and general shielding devices for this regime will be dealt with first.

Helium is clearly the best of the common shielding gases for performance and high quality $CO_2$ laser welding because it has the highest ionisation potential, which basically means it can absorb more energy before breaking down and promoting an unacceptable plasma formation. Therefore for a given welding condition higher weld penetration depths can generally be achieved. It is also an excellent weld shielding gas, plus being lighter than air it does not hang in the welding zone too long and degrade. Helium is recommended for $CO_2$ laser welding where high quality welds are required and is especially recommended for welding stainless steel and those materials such as titanium and zirconium where weld quality would be severely degraded by oxidation. The drawback to using helium in the UK and most of Europe is the high cost. In the UK, helium costs approximately 2.5 times that of argon and 10 times that of carbon dioxide and nitrogen.

Nitrogen (OFN) performs nearly as well as helium in terms of preventing an unacceptable plasma formation, but from a weld quality standpoint it can cause weld embrittlement in certain steels and should be treated with caution if considered for making welds from which very high quality and reliability are required. Nevertheless, acceptable weld quality for many applications is achieved with nitrogen which is perhaps the most common shielding gas for welding sheet steel automotive and domestic appliance parts. Interestingly, the laser welded food mixer whisk described in Chapter 3 was originally production welded using argon and a change to nitrogen was made to reduce production costs. Despite the numerous applications where nitrogen is used none are known to the author where the weld penetration is much deeper than 4 mm. This appears to be due to the marginally deeper penetration achieved with helium making more efficient use of the available laser power, or because deeper penetrations are usually used in more critical applications and weld embrittlement may be seen as a potential problem.

Shielding with carbon dioxide is not recommended for making welds using $CO_2$ lasers operated in the continuous wave mode. This is because carbon dioxide very quickly reacts with the focused laser beam, degrades and forms an unacceptable plasma cloud. The cloud is extremely brilliant and can radiate the laser light back on to and damage the focusing optic. Accompanied with this a rapid loss of weld penetration can occur within a very short weld length. Researchers at the Edison Welding Institute in the USA,[2] have reported good weld penetration using carbon dioxide to make laser spot welds in sheet steel when using a $CO_2$ laser operated in the pulsed mode. The pulsed operation and short welding duration prevent sufficient time for an unacceptable plasma formation, thus allowing an acceptable weld to be achieved.

Argon is an extremely good cover gas for preventing oxidation. It does not react and form an unacceptable plasma cloud quite as quickly as carbon dioxide, but in this respect is not as good as helium and nitrogen. When used with $CO_2$ lasers operated in the continuous wave mode and under circumstances where the heavier than air argon does not readily escape from the weld zone, it stagnates and suddenly, often within a distance of 100 mm from the start of the weld, forms a very dense plasma, causing a sudden loss of weld penetration. This occurs in situations where the gas is not easily pushed away by the incoming fresh gas from the shielding device. Typical situations are when welding in the down hand position, when well in from the edge on a flat sheet in restricted corners and in troughs. As a keyhole shield for long continuous welds, argon is only recommended where it can fall quickly away from the weld zone under gravity, or by extraction. Nevertheless, argon is useful when using a compound shield to weld thick sections where a cover gas trailing the weld keyhole is required to protect the weld bead during the final stages of solidification and cooling; a situation where plasma formation is not a problem. The use of argon in this instance provides substantial cost savings when compared with using helium for making long welds and where large production runs are required.

Mixtures of helium-argon, helium-nitrogen and argon-oxygen have been experimented with when laser welding high strength low alloy (HSLA) sheet steel at The Welding Institute in the UK.[3] The results show that although helium-nitrogen mixtures with nitrogen contents of 30 and 50% are cheaper than pure helium, the welding performance is no better than for nitrogen (OFN) alone. The results are summarised in Table 5.1. Helium-argon and argon-oxygen mixtures have poor performance as keyhole cover gases because they are prone to the formation of unacceptable plasma clouds.

*Table 5.1* **Comparison of best test results achieved with the shielding gases examined, which gave full penetration when laser welding lap seam joints in 0.7 mm HSLA sheet steel[3]**

| Shielding gas | Approximate UK cost as % of He cost | Nozzle diameter, mm | Gas flow rate, l/min | Weld width at joint, mm | HAZ width, mm | | Porosity, number of pores/ 250mm weld length | Hardness, HV5 | | Strength (average), N/mm weld length |
|---|---|---|---|---|---|---|---|---|---|---|
| | | | | | Top | Bottom | | Weld | HAZ | |
| He | 100 | 6 | 10 | 0.7 | 2.0 | 1.0 | 0 | 345 | 275 | 300 |
| $N_2$ | 10 | 6 | 10 | 0.7 | 1.9 | 1.1 | 7 | 370 | 290 | 315 |
| He-30%$N_2$ | 80 | 4 | 20 | 0.6 | 1.9 | 1.0 | 68 | 320 | 230 | 280 |
| He-50%$N_2$ | 60 | 6 | 20 | 0.5 | 1.9 | 1.0 | 42 | 370 | 280 | 345 |

Conditions: 2.5 kW, 2 m/min

Work on shielding gases for laser welding continues by gas suppliers and one should keep a watching brief on this area.

### Gas flow rates (Nd:YAG and $CO_2$ lasers)

For mass production the ideal choice of shielding gas, for a given application, is the cheapest in the minimum quantity which will consistently produce welds which meet their service requirement. Consequently, the gas flow rate is an important factor in terms of weld quality and cost.

The necessary gas flow rate is mainly dependent on the laser power and the access to and confines of the joint. When welding flat sheet steel and other easily accessed component joints at powers up to 3 kW, gas flow rates between 10 and 20 l/min generally suffice when using a correctly positioned co-axial or side tube shielding device (these devices are described shortly). At powers from 3 to 5 kW, rates of 15 to 30 l/min and between 5 and 10 kW rates of 25 to 40 l/min are suggested. It should be noted that these flow rates for steel will not prevent oxidation of the weld bead which occurs through residual heat after bead formation. If bright weld beads are required, gas shields which also trail the weld keyhole are necessary and usually together with higher gas flow rates. When welding materials sensitive to oxidation, such as titanium and zirconium and also where difficult access to joints in confined spaces exists, some experimentation with gas flow rates may be necessary. This also applies to welding applications where very high weld quality is required, such as specific nuclear and aerospace components. Under such circumstances weld underbead gas shielding may also be required and, for this, similar gas flow rates as for the top weld bead are suggested. However, where underbead shielding is used, care must be taken to ensure that the underbead gas does not try to exhaust through the keyhole and cause the entrapment of gas pores in the weld.

As the shielding gas flow rate is important from both weld quality and economy standpoints, care should be taken in selecting the correct gas flow meter. For accuracy a meter should be selected where the required flow rate falls within the central one-third of the meter scale. Care must also be taken to ensure that the meter scale and float are calibrated for the correct shielding gas.

## Gas shielding devices

### *Nd:YAG lasers*

For Nd:YAG laser welding, simple side tube shielding devices as shown in Fig. 5.5 are generally used, especially where precision placed spot welds are required. This is because the side tube provides good visual access to the target area for the spot weld. When making continuous seam and butt welds an annular skirt shield co-axial with the laser beam, Fig. 5.6, gives reliable weld protection and its alignment is not as critical as for a side tube and therefore perhaps better than the side tube for uni-directional welding in the downhand position. Co-axial nozzle shielding devices, Fig. 5.7, are however, more practical where the laser gun is manipulated by robot. They also provide the optic cover slide with some protection from possible weld spatter (see Chapter 11), because the force of the co-axial gas stream will partially oppose any particles travelling up the beam path.

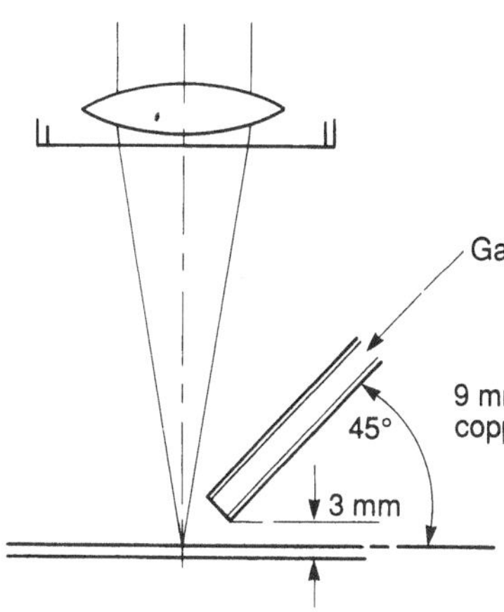


**5.5** A simple side tube gas shielding device suitable for Nd:YAG and low power $CO_2$ laser welding.

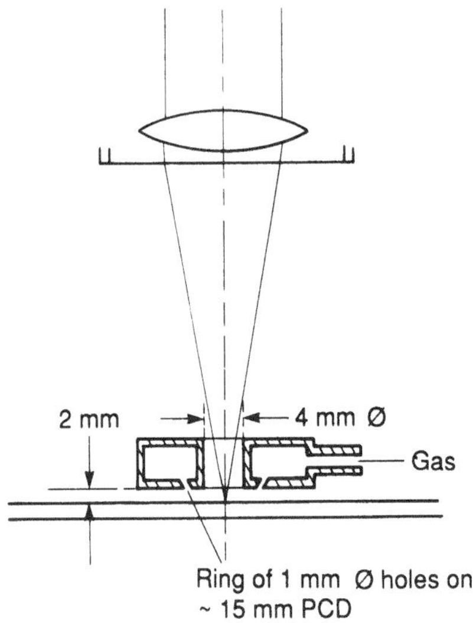


**5.6** An annular skirt gas shielding device suitable for Nd:YAG laser welding (originated from Ref. 4).

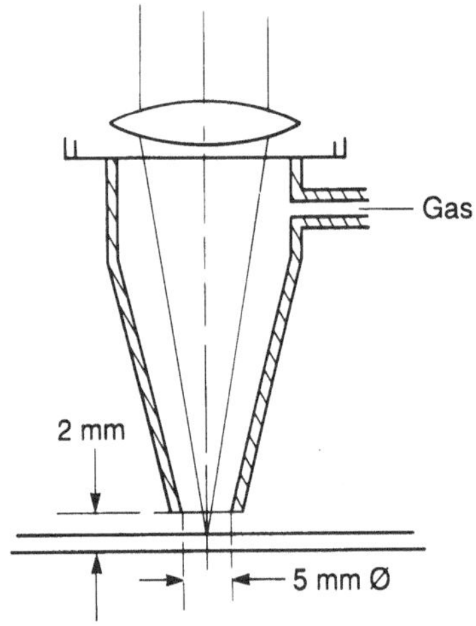


**5.7** A co-axial nozzle shielding device suitable for Nd:YAG laser welding (originated from Ref. 4).

*$CO_2$ lasers*

The advantages of the co-axial nozzle device, described for Nd:YAG lasers, also apply to $CO_2$ lasers worked at power levels up to and slightly above 5 kW. Much above this level, however, the possibility arises of an unacceptable plasma cloud forming within the nozzle mouth and destroying the focus ability of the laser beam. When this problem occurs it can be reduced to enable a small increase in laser power, by combining a side tube device.

A simple design for a detachable co-axial nozzle device for $CO_2$ laser welding is given in Fig. 5.8. Such a device is suitable for fast welding (more than 1 m/min) steel joints up to about 4 mm deep, with powers to around 5 kW, that is provided the joint configuration does not deny access to the nozzle and its stand off from the work surface can be kept to approximately 6 mm. The nozzle orifice diameter is not critical and for most applications 6 to 20 mm diameter, used in conjunction with gas flow rates from 10 to 40 l/min, are quite adequate for providing an acceptable weld cover shield. Nevertheless, it should be noted that orifices of 4 mm and less can cause weld porosity at these flow rates. This appears to be because of the increased concentration of gas and the resultant high gas velocity permitting gas to enter the weld keyhole and become entrapped.

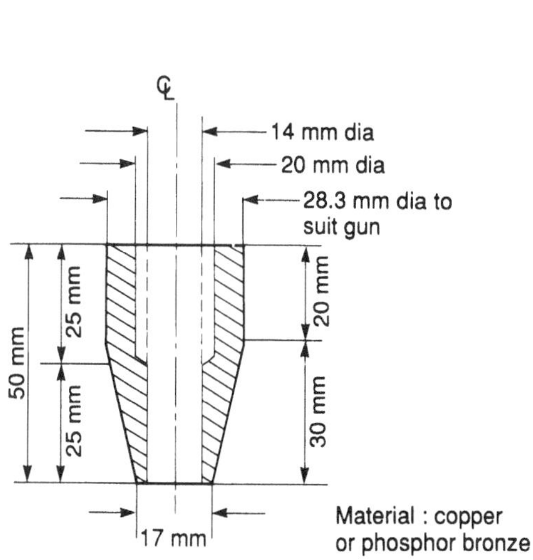


**5.8** Detail for a detachable gas shielding nozzle, which would be attached to a housing co-axial with the laser beam and similar to Fig. 5.7.

As laser weld penetration depths in steel are increased beyond 4 mm the weld cooling rate gradually becomes more prolonged, especially as welding speed is reduced, and the weld does not cool rapidly enough for a single nozzle device, whether side or co-axial, to provide adequate shielding of the weld bead before it leaves the area encompassed by the supply of cover gas. In such circumstances an extended shield which will also provide a trailing gas cover is necessary to avoid oxide scale in the surface of the weld bead. For welding speeds above 1 m/min, an extended shield can be achieved using a side tube device with a fish tail end, Fig. 5.9 or indeed a tandem tube device with the first

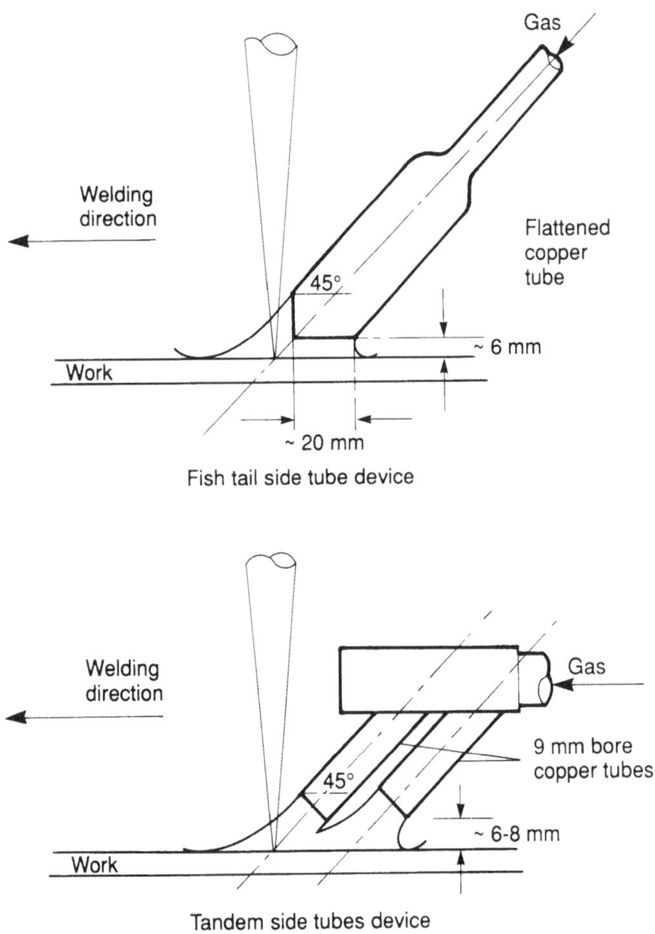


**5.9** Simple gas shielding devices for providing a short trailing gas cover.

tube aimed at the keyhole. These two devices are also useful for fast welding at laser powers above 5 kW since they eliminate the catchment point for a plasma cloud, which could occur in the mouth of a co-axial nozzle. However, when the thickness of steel exceeds about 7 mm even longer trailing shields become necessary if unacceptable oxidation of the weld bead is to be avoided.

Although the gas shielding devices for $CO_2$ lasers described so far are quite adequate for the general welding of steel, they are not recommended for welding those metals which are highly prone to weld degradation through oxidation or where bright weld beads are required for aesthetic reasons. For these applications, more comprehensive devices which provide a fore and aft gas shield are necessary. These devices can be gas filled enclosures, if the parts to be welded are small or specially shaped to encompass the weld area; some experimentation is usually necessary. Two devices which have produced good gas shielding results are shown schematically in Fig. 5.10 and 5.11. The first for making butt welds in flat sheet can be fitted with rubber side skirts to reduce the risk of entrainment of air. The second device was made and successfully used for gas shielding laser welded T butt joints in aero engine material C263 and a titanium alloy.

The emphasis so far has been on shielding devices for $CO_2$ laser welding at speeds above 1 m/min. When welding material thicknesses where the laser power only permits lower welding speeds, serious plasma formation occurs which reduces weld penetration depth, unless a plasma suppression device is used.

### Plasma suppression

When $CO_2$ laser welding in the continuous wave mode at speeds of less than 1 m/min, the longer laser beam/material interaction times lead to increased vapour or plasma ejection from the weld keyhole. This reacts with the shielding gas and, if not rapidly removed, forms an extremely dense plasma cloud just above the keyhole. The cloud which appears as a small very brilliant blue ball, a few millimetres in diameter, is not completely transparent to the laser light and re-radiates some of the laser beam in all directions before it can come to focus. This results in a general spreading of power before it reaches the material surface, thus widening the weld fusion zone close to the surface and reducing the weld penetration. The simple remedy is to use a higher laser power and increase the welding speed, but this of course is not always possible

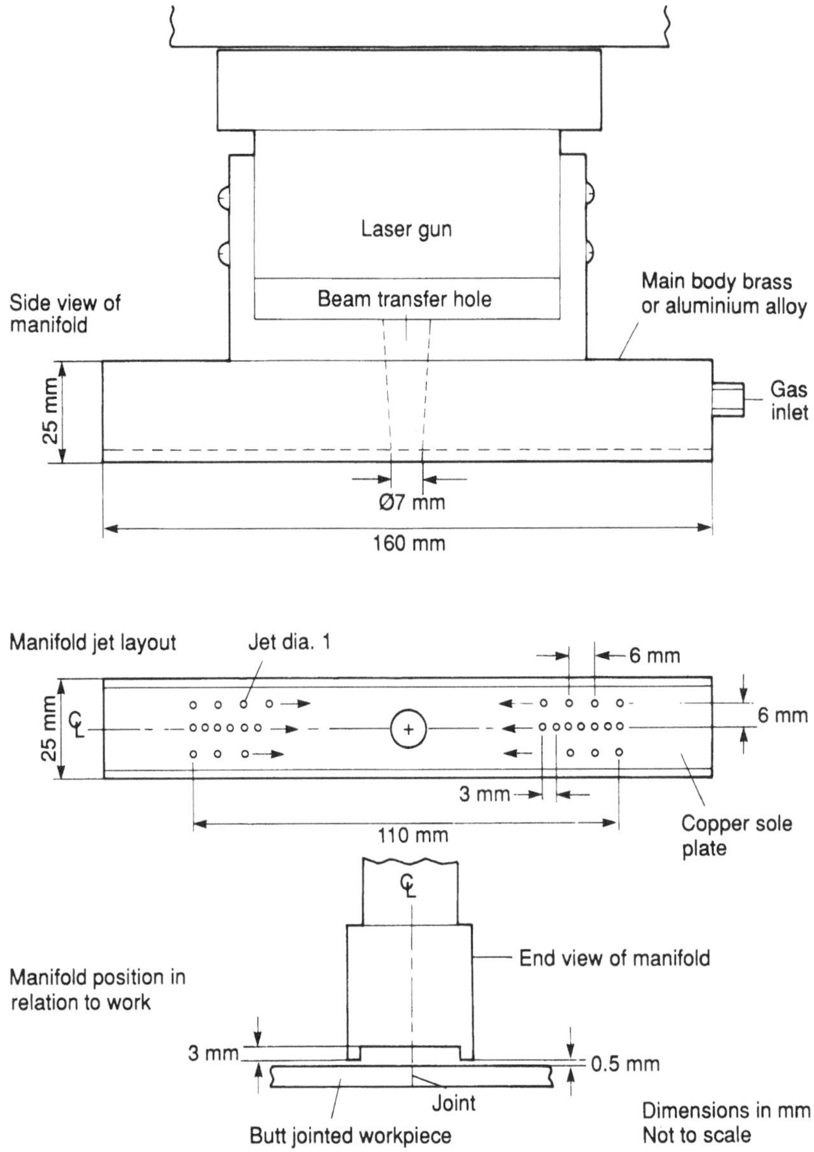


**5.10** Gas shielding device suitable for welding plain butt joints in materials prone to oxidation.

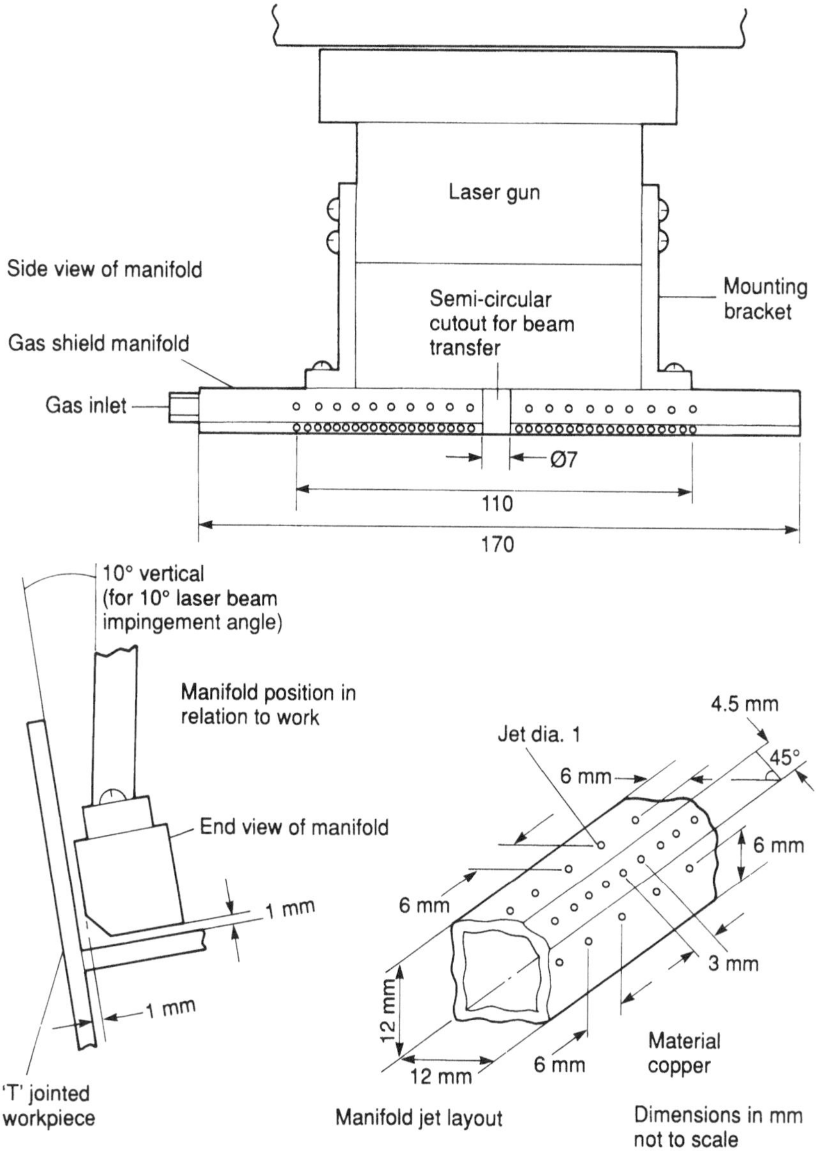


**5.11** Gas shielding device suitable for welding T butt joints in materials prone to oxidation.

where thick sections are required to be welded with a limited laser power. Consequently an alternative action becomes necessary and this is realised by the introduction of a plasma suppression jet into the gas shielding arrangement. The effect of this is clearly evident in Fig. 5.12 which shows sections taken through melt runs made with and without plasma suppression in 12 mm thick steel at various speeds from 0.5 to 1.25 m/min, using a laser power of 7.4 kW at the workpieces.

The plasma suppression jet is a small nozzle angled at 45° and aimed to blow a continuous helium gas stream at the plasma cloud both to remove the cloud and keep the area free of otherwise quickly degraded gas which would react and enhance plasma formation. The alignment of the gas jet stream with the laser beam and maximum plasma interaction point needs to be precise and therefore the nozzle is usually incorporated in the trailing gas shield manifold which forms a sturdy platform. The arrangement is shown schematically in Fig. 5.13.

The mounting arrangement of a combined suppression and trailing gas shield device, which is proven for laser powers from 5 to 10 kW is shown in Fig. 5.14. A general arrangement drawing of the device, which also depicts the X-Y-Z adjusters for accurately positioning the suppression jet in relation to the laser beam/plasma cloud, is given in Fig. 5.15.

The impingement point of the gas jet stream with the laser beam, the nozzle tip distance from the keyhole and the gas flow rate are important and need to be optimised according to how tightly the laser beam is focused. Experience has shown that tolerances are tighter where highly focused low order mode laser beams, such as from fast axial flow lasers, are used. This is thought to be because the increased intensity causes increased vaporisation and the formation of a keyhole with a very small opening from which a finer high velocity plasma stream exits and which requires a very accurately positioned suppression gas stream and flow rate to strike and disrupt it. Figure 5.16 provides suggested working conditions for the device design given in Fig. 5.15 which is suitable for focus spot sizes of 0.5 to 0.8 mm as typically produced by a 10 kW cross flow laser. For a 5 kW fast axial flow laser the suppression jet stream impingement position X and tolerance remain the same but the Y tolerance is reduced. Also a smaller jet diameter, 1.2 to 1.5 mm, and lower gas flow, typically 10 l/min rates are necessary.

To set the plasma suppression jet gas stream in relation to the focused laser beam, one must first find the focus position in relation to

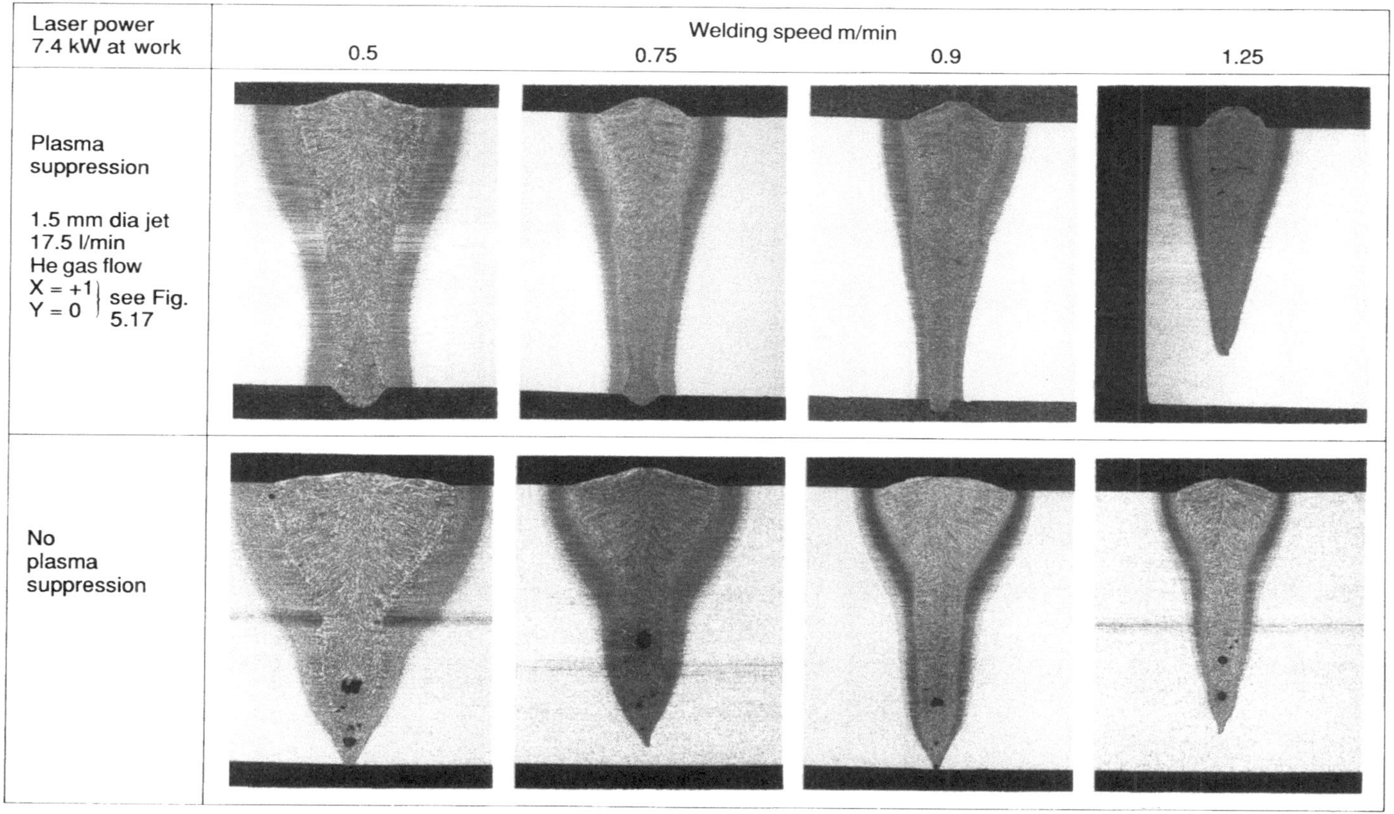


**5.12** The effect of plasma suppression on weld shape and depth when welding 12 mm thick steel at different welding speeds.

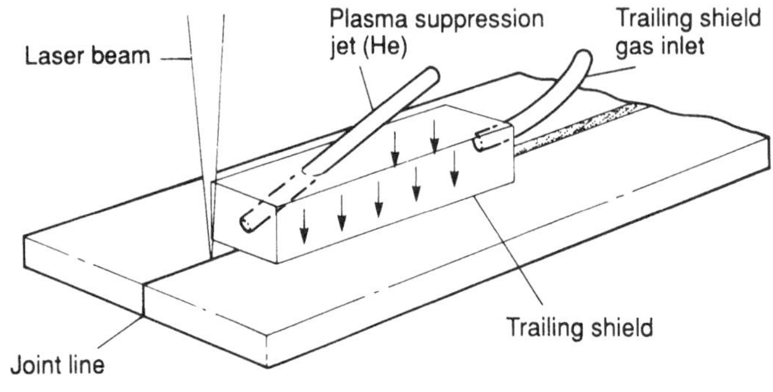


**5.13** Schematic arrangement of a combined plasma suppression device and a trailing gas shield used for $CO_2$ laser welding.

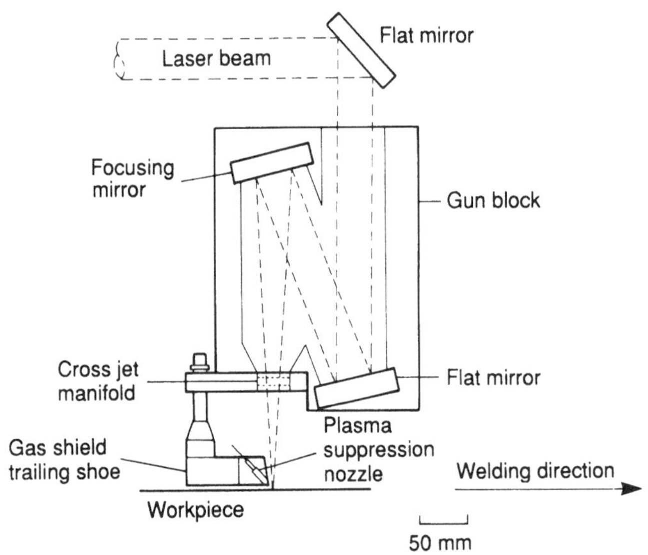


**5.14** The mounting arrangement of a combined plasma suppression and trailing gas shield device.

the work surface as described earlier in this chapter. Having defined the focus position the following technique is suggested to position the nozzle jet accurately. Clamp a piece of flat steel plate on the stationary work table with its upper surface at the beam focus position. Then at a low power setting (2–3 kW) open the laser beam shutter for a fraction of a second to produce a small melt spot on the plate surface. The position of the visible beam from the helium-neon alignment laser, which

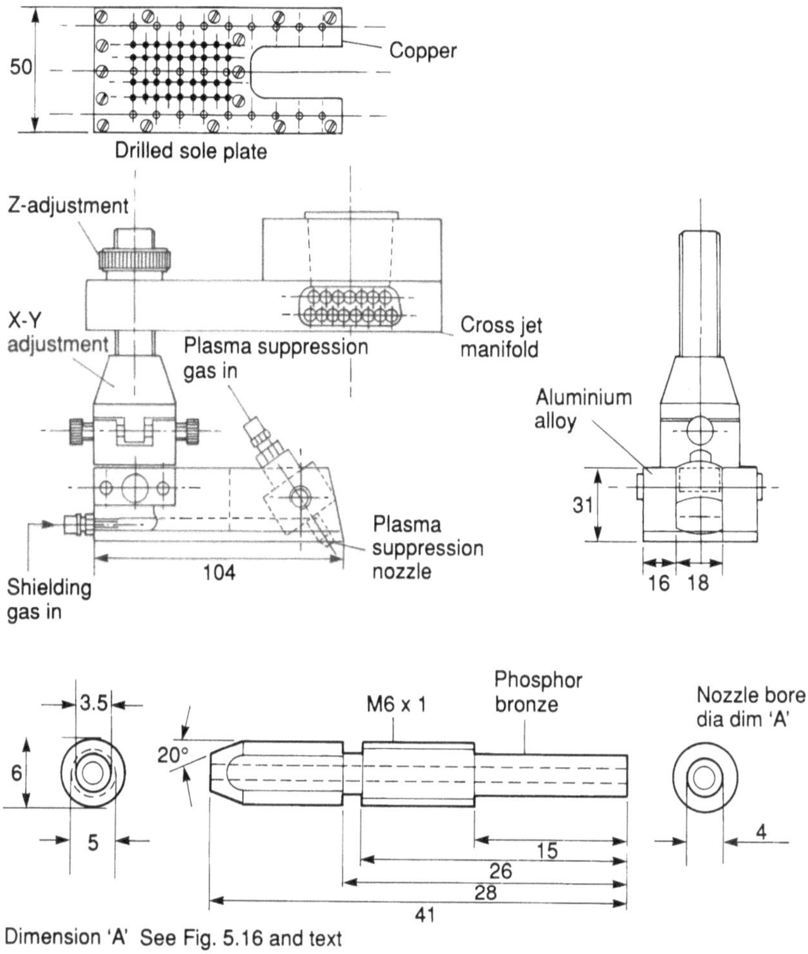


**5.15** General dimensions of a combined plasma suppression and trailing gas shield device.

should be on the same axis as the $CO_2$ laser beam, can now be adjusted if necessary to coincide with the centre of the melt spot. A twist drill, the same diameter as the bore of the plasma suppression jet nozzle being used, can be inserted in the nozzle and used to simulate the suppression gas stream. The settings shown in Fig. 5.16 can now be achieved by adjusting the X, Y and Z controls on the plasma suppression device.

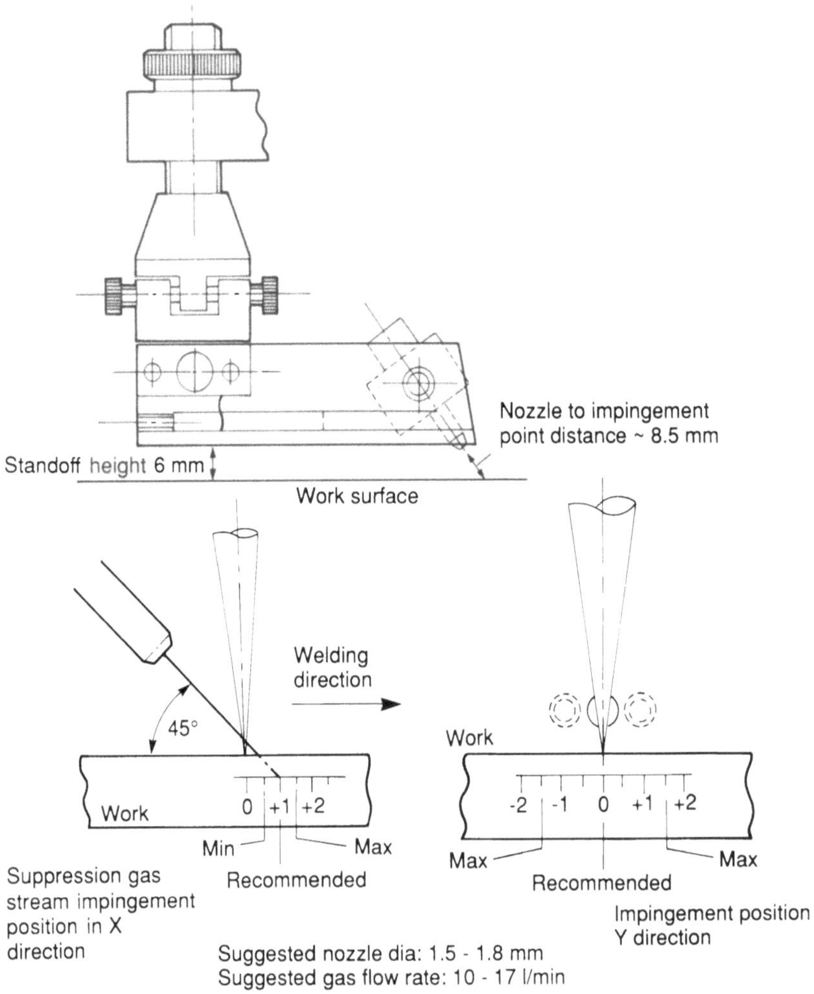


**5.16** Suggested working conditions for the plasma suppression device detailed in Fig. 5.15, when used at laser powers up to 10 kW and focused spot diameters between 0.5 and 0.8 mm.

## Laser power and welding speed

Having discussed the necessary practical steps for focusing the laser beam to achieve good all-round welding performance and how to enhance it by gas shielding, attention can now be turned to the effects of laser power and welding speed. The effect of these two parameters, when making continuous butt, seam or fillet welds, is easier to understand for $CO_2$ lasers operated in the continuous wave mode than for pulsed Nd:YAG lasers. This is because the number of joules/pulse on its own does not signify the rate of actual heat input since it is dependent on pulse width and amplitude. Furthermore, the welding speed can depend upon the pulse frequency and the percentage overlap of each spot. Therefore, the influence of power and speed on $CO_2$ laser welding is discussed first.

### Establishing conditions for $CO_2$ lasers

The average $CO_2$ laser operated in the continuous wave mode and welding steel, will achieve a weld penetration depth of 1.5 mm for every kilowatt of power (data based on powers up to 10 kW only) available at the workpiece when used in conjunction with a welding speed of 1 m/min, focusing optic f numbers between 6–9 and a suitable gas shield. Moreover, the relationship is almost inversely proportional as the weld speed is increased up to at least 5 m/min. Accordingly the graphs shown in Fig. 5.17 and 5.18 can be used as a basic guide for the selection of the advised laser power starting point for a required welding speed and penetration depth or vice versa. (They are also suitable for most other materials suited to $CO_2$ laser welding.) When using these graphs it must be remembered for reasons explained earlier in this chapter, that lower f numbers used in conjunction with a given power increase the welding speed for a given penetration, but the welds will be more narrow whereas with higher f numbers the opposite occurs.

The chosen combination of laser power and welding speed finally determines the weld energy (J/mm of weld length) and hence the weld properties and shape. Although the graphs in Fig. 5.17 and 5.18 give a guide to the nominal power necessary for a given weld penetration depth and welding speed, the weld width may prove to be unsuitable. $CO_2$ laser welds are rarely not narrow enough, however, occasionally they are not wide enough. Fortunately when making fully penetrating butt welds substantial increases in laser energy can be made without the risk of producing weld slumping (the molten metal falling out of the joint under gravity) or making blowholes and consequently weld width can be increased, often without changing the f number, by increasing laser power or decreasing welding speed.

From the above it will be apparent that some experimentation to establish the effects of laser power and welding speed on weld formation is necessary, if only to confirm the required weld depth will reliably be

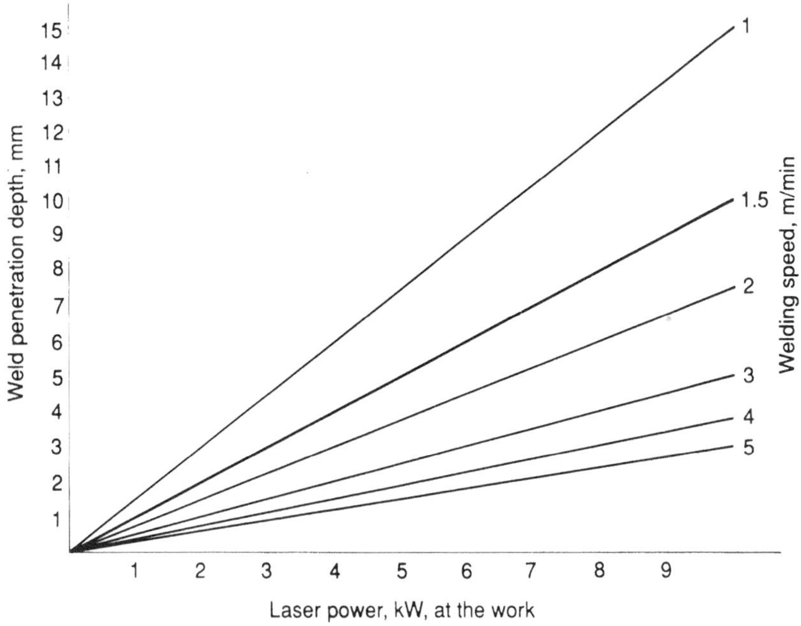


5.17 Approximate weld penetration depths that can be achieved in steel at different powers for given speeds, when $CO_2$ laser welding using focusing f numbers 6-9.

achieved within an acceptable parameter tolerance band. An easy method is to produce a weldability envelope which will also provide the tolerancing data. The envelope is established by making melt runs in test coupons and following a simple procedure. The actual components to be welded can be used instead of test coupons if they are of low commercial value, however, if only a partial penetration weld is required, each component will have to be sectioned before each step in preparing the weldability envelope can be taken. Test coupons should be the thickness of the required weld penetration depth and a practical length and width (75 × 25 mm is a useful size).

The procedure for producing a weldability envelope is first to select the laser power which should produce the required weld penetration depth and welding speed. For an example we will choose from the chart in Fig. 5.17, 4 kW for a 3 mm weld penetration at a welding speed of 2 m/min. Using these conditions as co-ordinates draw a power versus speed box matrix with the power and speed varied in + and − 10% steps as shown in Fig. 5.19. At the selected power and speed (4 kW at 2

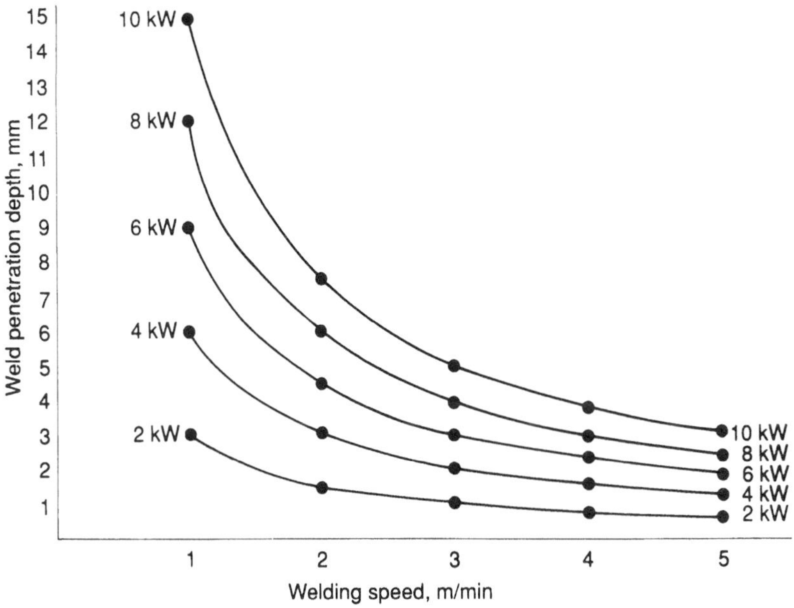


**5.18** Approximate weld penetration depths that can be achieved in steel at different speeds for given powers, when $CO_2$ laser welding using focusing f numbers 6-9.

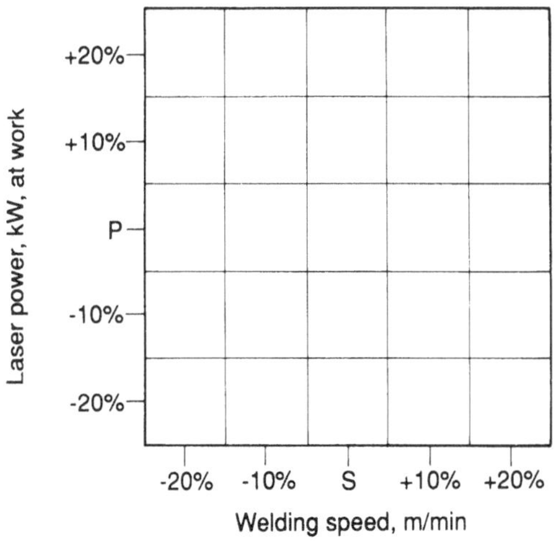


**5.19** Matrix for establishing power versus speed weldability envelope used for tolerancing conditions. Approximate P:power and S:speed settings can be chosen from Fig. 5.17 and 5.18 in relation to the required weld depth.

m/min) make a melt run in a test coupon and if on visual inspection it fully penetrates the coupon, place a tick in the respective box where the power and speed co-ordinates join. Then place a tick in all boxes vertically above and horizontally to the left, Fig. 5.20, where the power is 4 kW and above. On the other hand, if the melt run fails to penetrate, place a cross in the box and in all boxes vertically down and to the right where the power is 4 kW and below. The first melt run will thus isolate a substantial area in a corner of the matrix where penetration will or will not be achieved. If the melt run failed to penetrate then the power should be increased by 10% (or the speed decreased if extra power is not available) and the exercise repeated until penetration is achieved and the boxes completed accordingly. For our example we will assume that full penetration was achieved with the first melt run. In this case the second melt run should be made with the power reduced 10% (3.6 kW at the work). Here we will assume that on visual inspection full penetration is not achieved and thus the matrix can be further completed as shown also in Fig. 5.20. From the matrix so far we can see that full weld penetration depth can be achieved at the selected conditions of power and speed, but if the power falls a little below 4 kW an unacceptable penetration depth will occur. Alternatively, a power of 4.4 kW and a speed of 1.8 m/min will offer an all-round

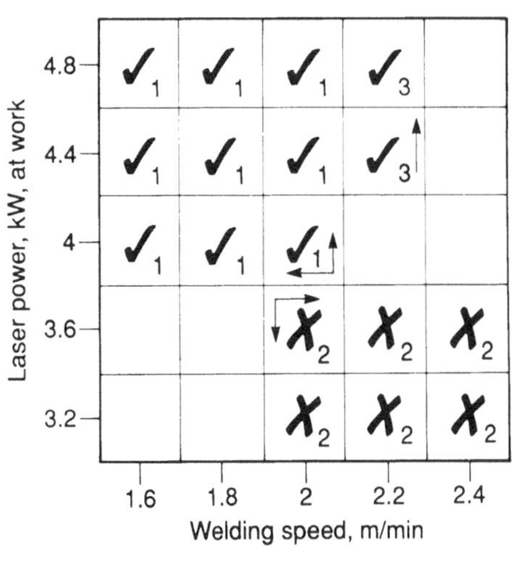


**5.20** An example of a practical test using the matrix system. It enables a weld power and speed tolerance band to be quickly ascertained by making only a few welds.

tolerance variation of 10%. To take the matrix one step further we will assume that a welding speed of 2 m/min with a tolerance of + 10% has to be achieved. From the matrix so far it is clear that 4 kW and a speed of 2.2 m/min is unlikely to produce the required penetration depth because of the reduction in energy, but by increasing the power to 4.4 kW at this speed the penetration depth is likely to be possible and therefore a third melt run should be made to establish this combination. If this is unsuccessful, which is unlikely from the data established so far, then a fourth melt run will be necessary with the power increased to 4.8 kW. Alternatively of course the results on the matrix could all be moved to the right by reducing the focusing f number, and repeating the exercise, that is providing the resultant reduction in weld width focusing and alignment tolerances, etc, can be accepted. Whatever the outcome, melt runs should finally be made and sectioned at those combinations of power and speed where the ticked boxes represent a practical working tolerance band. This action is necessary to ensure that the weld width and profile are acceptable and is particularly important when welding thick sections.

An alternative to the above is to make welds at different combinations of power and speed in practical increments for the range of power available. Then section each weld and plot the results against each power and speed combination. This approach is time consuming but worthwhile if a precise weld width to depth ratio is required.*

### Establishing conditions for Nd:YAG lasers

It is not possible to generalise on the weld depth that can be achieved for a given laser power and welding speed when it comes to pulsed Nd:YAG lasers. This is because different models of pulsed Nd:YAG lasers operate in different welding regimes which depend on their average power, pulse width and frequency ranges (pulse repetition rate). Consequently some lasers can only achieve a seam weld by producing a series of overlapping spot welds. Other pulsed Nd:YAG lasers can achieve a faster continuous seam weld by their ability to produce a continually moving molten weld pool or keyhole.

Nd:YAG lasers with output power levels below 500 W cannot usually achieve the power densities for keyhole welding and therefore the

*Melt runs do not truly represent butt joints. Conditions established by melt runs usually produce slightly more weld penetration depth when used on close fitting butt joints. However, the effect of the conditions can be lost if the fit-up of the joint and materials preparation is poor; these aspects are dealt with later.

welds are conduction limited (keyhole and conduction limited welding were described in Chapter 2). At low average powers, (less than 400 W) and the necessary related pulse lengths for welding (typically 4–8 ms) the penetration depth is usually limited to no more than the spot diameter produced. However, the spot diameters are usually in the order of 0.5–1 mm and the welds are ideal for spot welding thin materials; as in the case of the spot welded razor blade described in Chapter 3.

Conduction limited continuous welds are achieved by overlapping spot welds at relatively low repetition rates, which are necessary with low power lasers due to the long pulse lengths required. As a result each spot weld resolidifies before the pulse which produces the next overlapping spot arrives. To make hermetic welds in this manner, a spot weld overlap of approximately 70% is usually necessary. The spot overlap in conjunction with the pulse repetition rate and the effective spot diameter (the actual diameter produced by a single spot weld) determine the possible welding speed, which is calculated with the following formula:

Welding speed, mm/min = (Effective spot diameter, mm – Overlap length) × Repetition rate, Hz × 60.

At average powers between approximately 400 and 600 W and in conjunction with short pulse lengths, typically <5 ms and high repetition rates, the weld heat can be raised to the level at which solidification does not occur between pulses and a conduction limited continuously moving molten pool is maintained and consequently the welding speed is not fixed anymore in relation to the latter formula. The new generation of high power Nd:YAG lasers capitalise on this fact and with their high average powers, typically greater than 800 W, short pulse times, typically less then 2 ms and high pulse repetition rates, up to 500 Hz, they can produce and maintain a moving keyhole through the work with a high aspect ratio of depth to width. However, at this power level deeper welds, with a reduced aspect ratio, would be achieved at longer pulse lengths and repetition rates above 25 Hz.

There is a trend in weld shape formation which occurs as pulse width and repetition rate are adjusted in relation to laser power. The effect is shown diagrammatically in Fig. 5.21. The diagram also shows weld energy (joules), the pulse energy actually released by the laser which is measured by taking a small sample of the weld energy as it leaves the laser. This is a useful feature in that a given flash lamp power setting (an input power to the laser) will not always give the

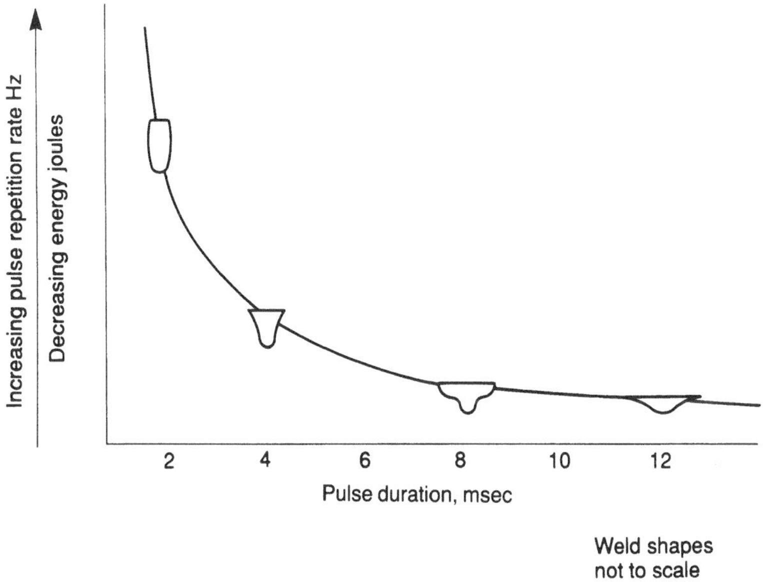


**5.21** The trend in weld shape formation in relation to pulse width and repetition rate when welding with an Nd:YAG laser.

same energy as the flash lamps, reflectors and laser rod age and therefore by monitoring the energy the power level can be corrected to maintain the required energy output. Energy values on their own however, have little meaning and must always be related to a pulse width and pulse shape if this control facility is available.

Pulse shaping is a more recent facility on Nd:YAG lasers and some lasers offer considerable control to encompass drilling and cutting. From a welding standpoint pulse shaping allows use of a pre-pulse. This can be in the form of a very short pulse with a much higher amplitude (peak power) than the immediately following main pulse. Such a pulse train is useful for welding high reflectivity materials because it ensures quick coupling of the laser beam into the work. Conversely a long low amplitude pre-pulse preceding a higher amplitude main pulse can eliminate piercing when welding those thin foils which are good absorbers of laser light.

To establish a welding condition one should first decide on the weld penetration depth and shape required. With respect to weld depth a typical 250 W Nd:YAG laser will comfortably achieve a 1.5 mm penetration in steel and a 1 kW laser 5 mm and intermediate powers somewhere between the two. If spot welds are to be made these laser powers can be used as a guide to the size of laser required likewise, Fig. 5.21 can be used to indicate the pulse width necessary to give the required spot weld shape. On the other hand if continuous welds are required some experimentation with high powers will be necessary if high welding speed is paramount. Published information,[5] indicates a 0.5 mm weld depth can be achieved at more than 3 m/min at a pulse repetition rate of 500 Hz when using an average power of 1 kW.

To make deep and narrow welds at high speeds, short pulse widths are required. However, care should be taken when using short pulses ($<$1 msec) and high powers (typically 1 kW) as weld undercutting can occur through excessive vaporisation and material ejection.

When increasing the laser power the beam divergence angle also increases on most Nd:YAG lasers, this affects the focus position and incident spot size. Therefore, when increasing power it is usually also necessary to adjust the laser beam expansion telescope (Fig. 1.8, Chapter 1) to maximise the power density achievable by the focusing optic used.

## Material preparation prior to welding

Once welding parameters have been established their performance will depend on the preparations of the joint faces, joint fit-up and the accuracy to which the laser beam is aligned with the joint centreline. For high quality welding the joint faces should be free from any contamination such as grease, paint, dirt and oxide scale. If the parts are to be autogenously welded the joint faces should be smooth and in very close contact. On butt joints between thin sheets, such as those used in autobody construction, guillotined edges can be used provided they are square to the sheet surface and straight. The joint faces for thicker sections can be turned, milled, shaped or planed and, however produced, must make good contact with their mating part. These actions are necessary to avoid the two main fit-up problems: joint gaps and surface mismatch, which can occur together. Gaps are more serious when welding thin sheets than thick plates and both cases are dealt with in Chapters 6 and 7 respectively as is the importance of laser beam to joint alignment.

Chemical degreasing of parts such as washing with petroleum ether and Genclene, and vapour degreasing with trichlorethylene, should be conducted away from the welding zone and the laser beam transmission lines. This is because the vapours given off by these agents can impair transmission of the laser beam by causing thermal blooming, which is described in Chapter 11. Moreover, cleaning agents must be adequately rinsed off and dried before parts are welded, otherwise they may result in producing a high degree of weld spatter and porosity in the welds. Unremoved cleaning solutions containing sulphates and phosphates are particularly prone to these reactions.

## Welding performance troubleshooting checklist

As described in Chapter 4, welding performance is dependent on the material to be welded and some materials are welded more easily than others. Also coatings and variations in thicknesses can cause major changes and disrupt the welding operation. The following troubleshooting checklist ignores these facts and assumes that the welding conditions and parameter tolerances have been correctly established for a normally readily weldable material and that something has gone wrong.

| Symptom | Possible cause |
|---|---|
| Sudden loss of weld penetration. | Loss of power - check power setting and calibration. |
| | Welding speed accidentally increased - check setting and calibration. |
| | Focusing optic not functioning correctly - if the optic is a lens, check to see if it is cracked. If a reflective optic is used check the incident beam is on the optical axis (Chapter 11). |
| | The workpiece is out of focus - check the work table or focusing optic have not been moved. |
| A gradual loss of weld penetration over a period of days. | The focusing optic and/or the beam transmission mirrors are becoming dirty - clean. In the case of Nd:YAG lasers, check the focusing optic cover slide. |

| | |
|---|---|
| | The laser beam transmission system to the focusing optic has gradually moved out of line - check alignment (Chapter 11). |
| | Nd:YAG laser flash lamps are degrading or the lamp reflector is dirty - check, replace and clean as necessary. |
| During the welding operation weld penetration gradually decreases and the brightness of the plasma cloud above the keyhole increases. | The focus position is moving due to thermally induced movement of the workpiece - check workpiece clamping arrangement. |
| | An inadequate flow or use of the wrong shielding gas - check supply and flow rate. |
| | The focus position is moving through thermal lensing of either the focusing optic, beam transmission train or laser output window - check optics for cleanness and adequate water cooling. |
| Each time the welding operation is repeated, weld penetration suddenly decreases to a narrow bead of surface melting and the plasma cloud above the keyhole loses its brightness and turns yellow. | Sudden loss of power density - check for thermal blooming (Chapter 11) in the beam transmission system scattering the laser beam. A purging gas of dry nitrogen or air in the transmission system may be necessary to solve this problem. |
| | Thermal lensing of dirty laser beam transmission windows into glove boxes are also a common cause of this symptom. |
| Excessive weld spatter. | The laser power is too high - check power setting and calibration. |
| | The welding speed has decreased - check setting and calibration. |
| | The weld faces are contaminated with oil oxide paint, etc - check parts for cleanness. |

| | |
|---|---|
| | The shielding gas flow rate is too high and blowing out molten metal - check gas flow rate. |
| | The shielding gas nozzle has become blocked, thus increasing the gas velocity - check and clean. |
| | Excessive dross built up in the channel beneath the joint line of a butt welding jig is melting and ejecting through the weld keyhole - check and clean. |
| Weld blowholes on the surface and weld porosity. | Inadequate gas shielding - check gas, flow rate and shielding device position. |
| | Oxygen content of steel is too high (see Chapter 4) - check material specification has not changed, if necessary have chemical analysis taken to check. |
| | The weld faces are contaminated prior to welding - check and clean as necessary. |
| | The material contains elements prone to outgassing, e.g. zinc in brass - such materials are not suitable for laser welding. |
| | Excessive zinc coating on steel - see Chapter 4. |
| | Build up of soot and weld spatter on gas shielding device falling off and into weld zone - check and clean. |
| | Dross build-up, in channel beneath the joint line of butt welding jig, melting and being expelled into the weld keyhole - check and clean channel. |
| Weld cracking. | The material specification has changed - check especially for increases in carbon, sulphur and phosphorus levels (see Chapter 4 for effects). |

| | |
|---|---|
| | Part location and fit-up have changed removing the allowance for weld contraction stresses during solidification - check that joint design or dimension changes have not occurred, examine fit-up and clamping (joint designs to eliminate cracking are described in Chapter 6 for sheet and Chapter 7 for plate and tube). |
| | The weld has been made over a surface treatment such as carburised or sherardised (Chapter 4) - check masking of weld area is correct before surface treatment. |
| Weld slumping or undercutting. | Welding power increased or welding speed decreased - check settings and calibration. |
| | Poor part fit-up producing gaps at weld faces or reliefs and chamfers are too large - check part preparation and location for welding. |
| | Excessive gas flow rate or velocity - check setting and shielding gas nozzle for blockage. |

## References

1 Nakajima N et al: 'Fundamental study on 1 kW class YAG laser using optical fibre'. IIW DOC IV-451-88.
2 Chennat J C and Albright C E: 'Shielding gas effects in pulsed $CO_2$ laser spot welding'. LIA Vol 44, ICALEO, 1984, 76–85.
3 Dawes C J et al: 'Shielding gases for $CO_2$ laser welding - HSLA autobody steel'. Conf Proc. Sheet Metal Welding Conference IV, Oct 1990, American Welding Society publication.
4 Watson M N: 'Micro-spot welding with a pulsed Nd:YAG laser'. TWI Research Bulletin 1983, 24, 10.
5 Lumonics Ltd: 'JK 706 series laser data sheet' (printed May 1990). Lumonics Ltd, Cosford Lane, Swift Valley, Rugby, Warwickshire, CV21 1QN, UK.

Chapter 6

# Welding sheet metal parts

Ductile materials, typically up to 3 mm thick, can be easily formed into a range of joint configurations which are not always possible or practical in thicker material sections. Most sheet metal fabrications produced by laser welding are in materials less than 3 mm thick and are made at welding speeds in excess of 2 m/min and laser powers less than 5 kW. The main applications are found in the automotive, domestic appliance and electronic packaging industries. This chapter therefore concentrates on sheet metal welding data pertinent to these types of industries. The selection of laser power and welding speed is not dealt with here, since it was covered in the previous chapter.

In general the data presented are related to welding practice when working with the different sheet metal joint configurations. In some instances tolerance figures are presented as a percentage of the sheet thickness. Where this occurs, in this chapter, it must be remembered that the percentage value only applies for materials up to 3 mm thick. Some mechanical properties data are also presented where pertinent to the subject under discussion. However, these data are only presented as a guide since such results depend on the type of material and its surface coating, if present. Although certain surface coatings, such as zinc, can have a major effect on weld formation and quality they are not discussed here since the effect of coatings, together with surface treatments, are dealt with in Chapter 4.

## Joint configurations

Examples of sheet metal joint configurations which are suitable for laser welding are shown in Fig. 6.1. Joints 1, 2 and 4 are often used in both sheet and tubular components, whereas 3, 5 and 6 are generally confined to sheet metal parts. From a welding practice standpoint the lap fillet and T butt joints can be grouped together, as can the lap edge

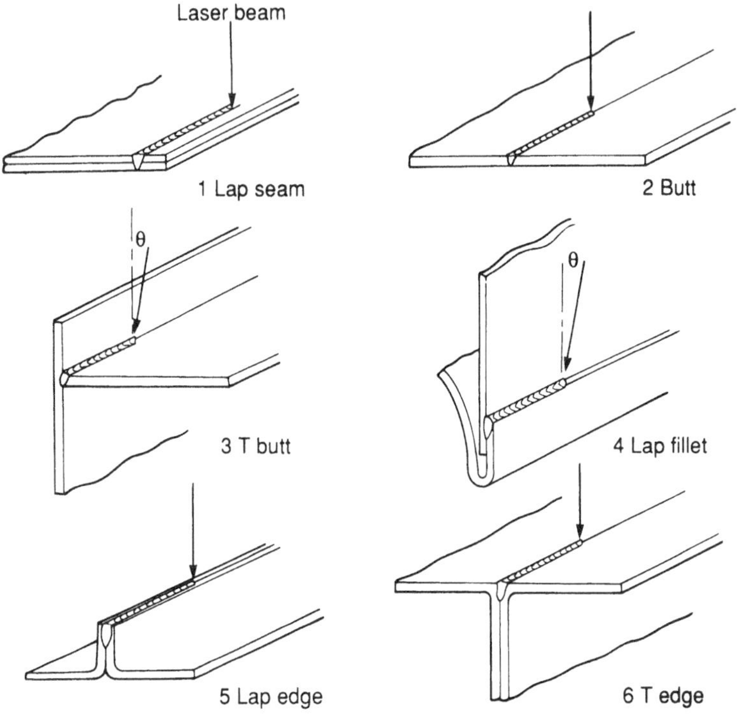


**6.1** The basic sheet metal joint configurations that can be laser welded.

(also known as a standing edge joint) and the T edge joints (the T edge joint is also often referred to as the coach or flared T joint).

### Lap seam joint

The lap seam joint is the most common laser welded sheet metal joint because it offers the most versatility from a design standpoint in respect of ease of manufacture under production conditions. Fortunately it is also the most simple to laser weld from the process application point of view; it does not present laser beam to joint alignment problems and joint preparation and location are not normally too difficult to achieve.

The disadvantages of the lap joint are that weld energy is expended in penetrating the first sheet thickness before the joint face is reached, the weld width is very narrow, usually less than 0.8 mm at the joint interface, and, unless the thicknesses of the sheets to be joined are fully penetrated, there is no visual indication that the weld fusion zone has penetrated adequately across the joint interface.

For lap joints to have good shear strength it is advantageous if the weld width equals the sheet thickness. Nevertheless, the narrow weld width can in some instances enable lap seam welds to be made in narrow flanges close to the sheet edge and thus allow material savings to be made.

Intimate lap joint fit-up is difficult to achieve in thin sheet materials especially where the joint follows a curved surface. Fortunately small joint gaps of less than 10% of the sheet thickness present no significant problems with respect to weld undercutting and peel strength. Gaps equal to 25% of the sheet thickness, as expected, promote serious undercutting and hence loss of peel strength due to the large reduction in material cross section caused by weld slumping (material drop through). The gap also affects weld keyholing and loss of penetration into the lower sheet can occur. These effects are shown in the lap joint macrosections of welds between 1 mm thick steel sheets in Fig. 6.2. To avoid unacceptable gaps, careful clamping of lap joints is important and can be achieved by individually loaded finger clamps if the joint surface profile is irregular, or by substantial flat clamping bars, Fig. 6.3, if the surface is flat.

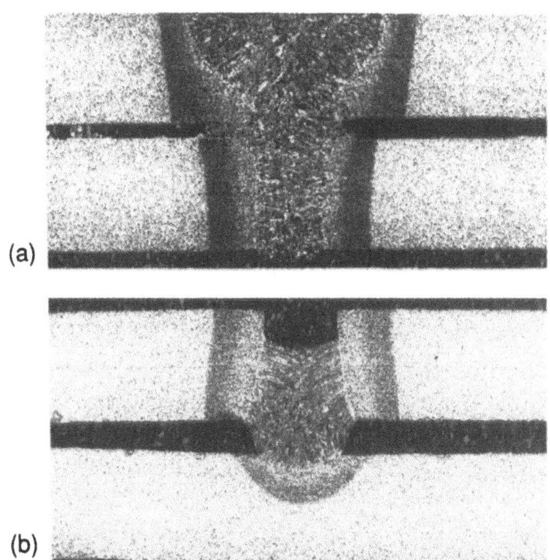


**6.2** Transverse cross sections through lap seam welds between 1 mm thick low carbon steel sheets, made with joint gaps of 0.1 mm (a) and 0.25 mm (b); note the weld slumping and loss of keyholing. Both welds were made at welding power, 3 kW and welding speed, 2 m/min.

6.3 Large flat clamping bars being used to hold lap joint faces in close contact. Note the two parallel linear lap seam welds which have been made in joining each corrugation to the flat sheets.

When welding zinc coated steel sheets it can be desirable to have a small gap (typically 0.1 mm) between the faces of lap joints to enhance weld quality and reproducibility; the reasons are discussed in Chapter 4. The small gap required can be achieved by the clamping arrangement shown in Fig. 6.4 which enables a uniform gap to be achieved without the need for a precise clamping force.

Good lap joint fit-up for welding pressed sheet material lids into tubes or deep drawn cans, can be achieved by careful attention to design detail. Figure 6.5 shows two common designs which are often used. The example shown in Fig. 6.5 (a), although employing less material, has to be carefully sized to obtain good fit-up, whereas that in Fig. 6.5 (b) can flex and thus self-adjust to match small variations in the tube or can body aperture. The latter design is recommended and is used on the lid of the fuel filter can shown in Fig. 6.6. End closures of this nature can be square or rectangular if a sufficient radius can be achieved at the corners.

Although continuous lap seam welds in sheet steel are very narrow (usually less than 0.8 mm) they have comparable impact properties (in 1 mm thick steels) to a line of 5 mm diameter resistance spot welds with a pitch spacing of approximately 20 mm. For example, experimental box sections[1] developed to represent energy absorbing car body structures (accident crush zones, which in practice are required to produce controlled collapse and therefore should not be prone to sudden

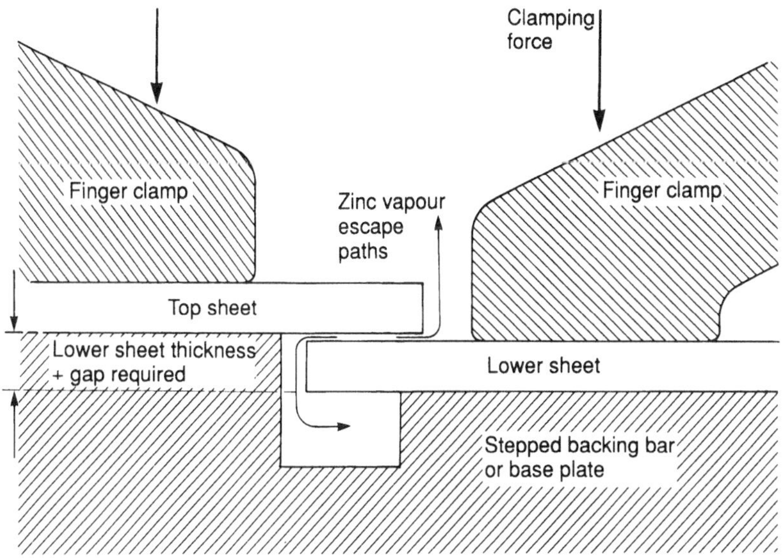


**6.4** A method of achieving a controlled gap between the faces of lap joints when welding zinc coated steels.

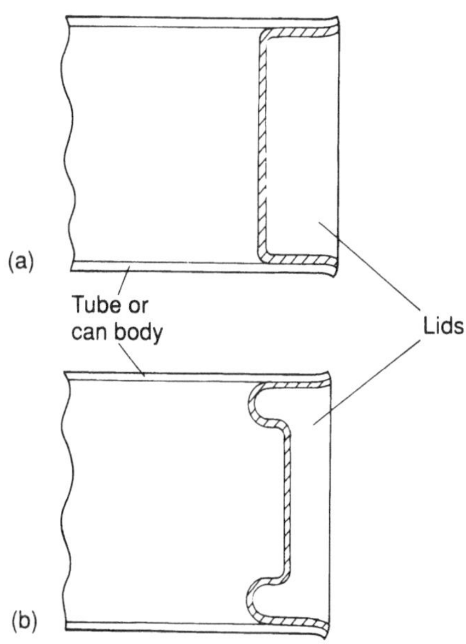


**6.5** Two common types of pressed sheet metal lids used to close tubes or cans. Type (a) has to be carefully sized to produce a close fitting lap joint, whereas type (b) is able to flex and self-adjust to achieve good fit-up and thus increase manufacturing tolerances.

**6.6** A laser welded lap seam joint between a fuel filter lid and can body, employing the lid design shown in Fig. 6.5(b).

weld failure on impact) have been laser seam welded and compared with resistance spot welded examples. The results of impact tests, using an energy of 6000 J, for both low carbon and formable high strength steels, show the impact energy necessary to deform the box sections by 1 mm was in both cases approximately 40 J for low carbon steels and 55 J for the high strength steels. Examples of laser welded box sections before and after impact testing are shown in Fig. 6.7.

The steel strength of a lap seam weld made in thin sheet material depends on the direction of peel and is greater in the direction transverse to that of welding. Where this presents a problem, the peel strength normally in the direction of welding can be improved by replacing the linear weld with a different pattern such as a sine wave or a displaced semi-circular wave form as shown in Fig. 6.8. When using these two weld patterns the welding speed remains unchanged, but the forward production speed of the work is of course reduced as the frequency and amplitude of the wave form is increased. Moreover, the higher the frequency the greater will be the heat input into the work as can be seen by the different heat bands in Fig. 6.8. To produce such

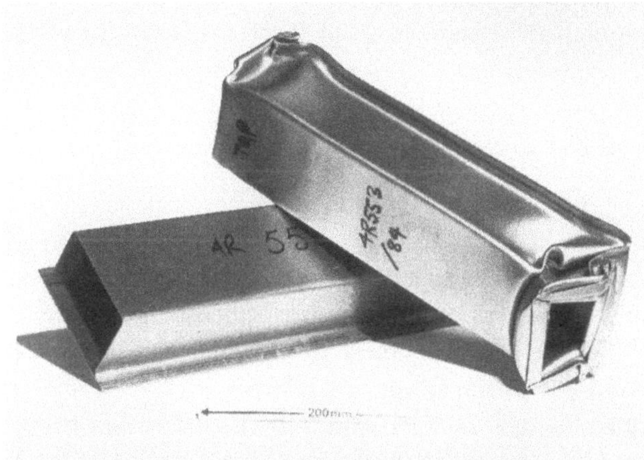


**6.7** Laser welded experimental box sections before and after impact testing; in this test the welds must not fail upon impact. To form the box sections continuous welds in lap seam joints have been made between 'top hat' formed sections and flat sheets.

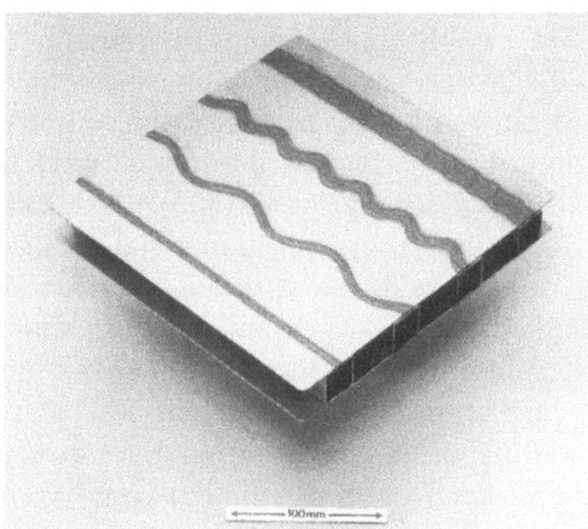


**6.8** Different lap seam weld patterns which can be used to enhance structural strength.

weld patterns requires either moving the work on a X-Y table or a laser beam manipulation system. Both of these facilities are described in Chapter 10.

### Butt joint

A butt joint in thin sheet steel is perhaps the most difficult of all the joints to make reliably by laser welding without special care, especially when making long joints. Apart from correctly established welding conditions, the two most important requirements are laser beam/joint alignment and good joint fit-up.

The distance or tolerance by which the focused laser beam can be displaced to one side of the joint line and still produce a weld zone which encompasses the entire joint is very small because laser welds are inherently very narrow. The absolute value is dictated by the minimum dimension of the fusion zone width produced by the welding conditions employed and for confidence during production a figure of no more than 40% of the minimum width is recommended. To establish this necessitates taking a transverse cross section through a weld. However, such a section would normally be taken when the welding condition is established; establishing welding conditions is described in Chapter 5. It is difficult to generalise on a laser beam/joint alignment tolerance figure, since, for a given power, the weld width will decrease as the welding speed is increased and vice versa. Furthermore, the materials being welded will have a major influence on the weld width formed. For example, for a given material thickness, laser power and welding speed, welds in stainless steel are generally marginally narrower than in mild steel, and welds in titanium are noticeably wider and more nearly parallel than in both of the latter. When welding sheet steels at speeds more than 2 m/min the minimum portion of the weld width appears to be just over twice the calculated focused spot size (see Chapter 5 for the calculation formula) and therefore if the beam alignment is found to be more than a distance equal to the calculated focused spot diameter to one side of the joint line, when welding steel, there may be reason for concern regarding whether the joint line has been encompassed by the weld. For example, Fig. 6.9 shows a transverse cross section through a weld butt joint made with a calculated focus spot size of approximately 0.2 mm diameter and with the centre of the focused spot displaced 0.5 mm to one side of the joint line. Externally this joint appeared acceptable but internally much of

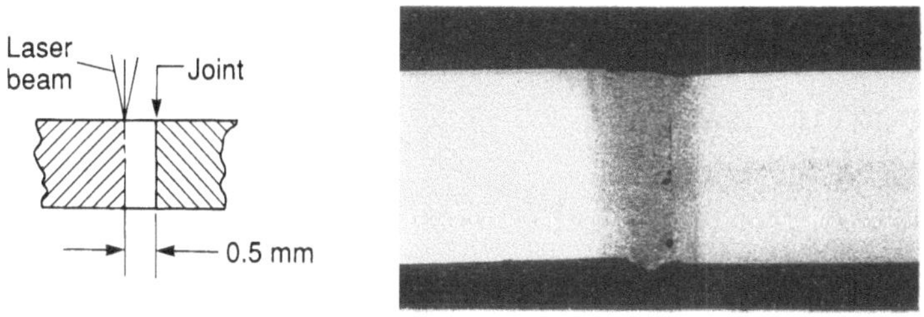


**6.9** A transverse cross section through a butt weld in 3 mm steel where the laser beam was offset 0.5 mm from the joint line. The fusion zone has failed to encompass much of the joint below the sheet surfaces.

the joint face is unwelded. Had the laser beam been displaced no more than 0.2 mm from the joint, the weld would have encompassed all the joint line. The importance of ultimately establishing laser beam/joint alignment tolerances by taking measurements from transverse cross sections of welds, made at the conditions to be used in production, is emphasized by this example. An increase in tolerance to laser beam/joint alignment and joint fit-up can be achieved by laser beam spinning which will be described shortly.

The level of care necessary to establish good butt joint fit-up increases with decreasing sheet thickness. When designing parts which require butt joints the following aspects need to be considered with respect to locating the parts for welding:

- Joint face fit-up (Fig. 6.10);
- Joint surface mismatch;
- The combination of both the above.

Joint face fit-up is the most critical aspect because most welds between sheet metal butt joints are made autogenously. The tolerance to gap between the joint faces depends firstly on the focused laser beam spot size used and secondly on the amount of weld undercut that can be accepted in service. In thin sheets, as used in automobile body work (typically 0.7 to 1 mm), face gaps should never be much greater than half the focused spot size and, since 0.2 mm is a typical spot diameter, gaps greater than 0.1 mm are not recommended. One of course could argue the case for reducing welding speed or increasing laser power and using a larger focused spot size (actions which could lose the advantage of laser welding over thin sheet arc welding techniques). However, as the gap or face mismatch increases, the volume of molten

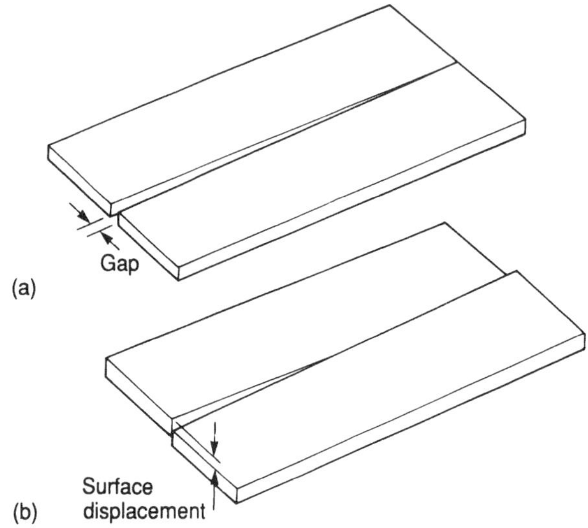


**6.10** Butt joint face mismatch (a) and surface mismatch (b) which require careful control when welded by power beam techniques.

metal produced to bridge the gap decreases resulting in undercutting, and, in extreme cases, voids, which cause a reduction in weld cross section and hence strength.

As the sheet material thickness increases, the face gap which can be tolerated *does not* increase proportionally (i.e. 0.1 mm for 1 mm *does not* mean 0.2 mm for 2 mm) as can be seen in Fig. 6.11, which shows a transverse microsection taken through butt joints in 2 mm thick steel which had face gaps of 0.25 and 0.12 mm prior to welding. In the author's experience, when welding with a calculated focus spot size of approximately 0.2 mm and where a weld undercut of up to 15% of the sheet thickness can be accepted, gaps of up to 0.12 mm in 2 mm, and up to 0.18 mm in 4 mm thick steel, can be used as a practical indicator provided the focus spot size is not less than 0.2 mm. The reason a gap so close to the calculated spot diameter can be used on 4 mm thick steel, without the beam almost completely passing through the joint, is because the laser beam will be converging or diverging, according to the minimum waist position of the focused beam about the work surface, and thus is absorbed by the walls of the gap.

In thin sheet materials less than 1 mm thick it is easy to achieve poor surface fit-up if the work is inadequately clamped and therefore special care is necessary if an unacceptable loss of joint cross section is to be

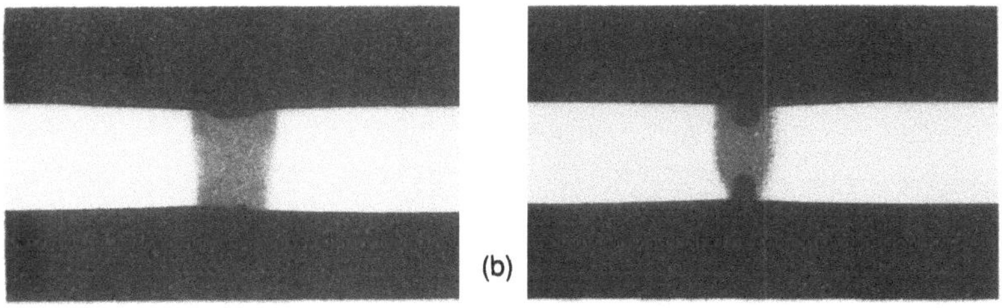

(a) (b)

**6.11** Sections through butt joints in 2 mm thick steel made with face gaps of 0.12 mm (a) and 0.25 mm (b). The total undercutting of the sheet surfaces is less than 15% of the sheet thickness in (a) whereas it is nearer 50% in (b). Both welds were made at welding power, 4 kW and welding speed, 2 m/min.

avoided. Good butt joint surface fit-up between flat sheets can be achieved by using robust clamping bars which have their clamping force equally distributed across each sheet and are placed relatively close to the joint; within 6 mm is typical. Where weld lengths in excess of 1 m are required, such as in coil joining or when fabricating large blanks for presswork, special purpose-built clamping equipment is necessary. Such equipment should also incorporate lateral clamping to eliminate joint opening, see Chapter 10.

Surface mismatch has little effect on weld formation between sheet metal butt joints providing the joint faces are well butted together. Figure 6.12 shows sections through test welds in 2 mm thick low carbon steel where the sheet surfaces have been deliberately displaced 50% of the sheet thickness (as in Fig. 6.10(b)). Despite the reduction caused in the thickness of the joint, no blowholes or slumping of the weld metal occurs in fully killed steels. Moreover, the part of the laser beam which is incident on the top edge on the upper of the two mismatched plates melts the edge down into the joint and forms a smooth fillet. This fillet can be an advantage where mismatch is unavoidable, as it helps limit the loss in cross sectional area and produces a smoother transition from one sheet to the other.

It is interesting to note from Fig. 6.12 that, although the laser beam was applied at 90° to the surface, the fusion zone produced was not in the same place as the beam where the specimen surfaces were mismatched. The fusion shape and direction is dictated by the heat sink effect produced by the change in joint geometry.

Finally, butt welding dissimilar sheet thicknesses generally presents few problems if the fit-up is good. Examples are discussed and shown in Chapter 3.

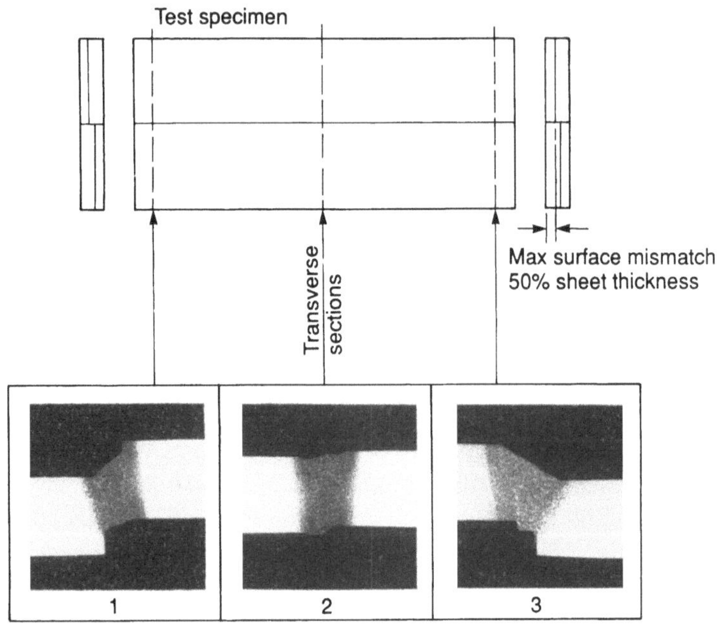


**6.12** Sections through butt joints made in 2 mm thick low carbon steel sheets, where the sheet surfaces have been displaced 50% of the sheet thickness and showing that surface mismatch can have little effect on weld formation.

### Lap edge and T edge joints

The laser beam/joint alignment and joint fit-up requirements for lap edge and T edge joints are basically the same as for the butt joint. However, when welding sheet thicknesses much less than 0.7 mm welding conditions should be chosen which restrict the fusion zone width, i.e. high welding speeds should be used. In the case of the lap edge joint, this is to avoid breakout by the fusion zone through the side wall of the joint, thus destroying the aesthetic appearance of the piece part. This of course also applies to the T edge joint, but here there is a more serious further effect, a wide fusion zone also causes sheet thinning to occur at the point where the two sheets flare apart. An example of sheet thinning at this position is shown in Fig. 6.13.

### T butt and lap fillet joints

For T butt and lap fillet joints it is necessary to apply the laser beam at an angle otherwise the converging beam will clip the projecting

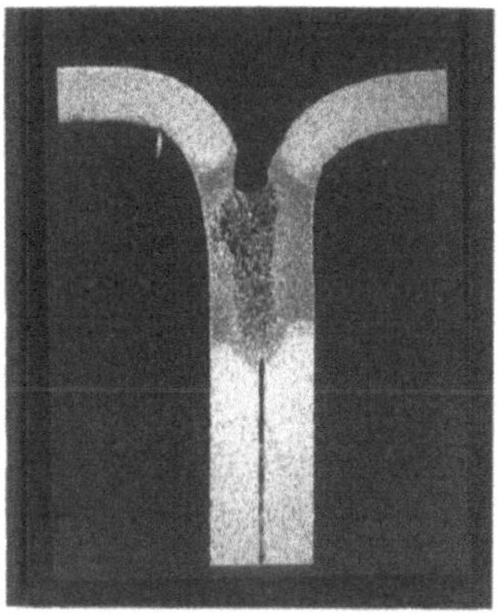

6.13 A section through a T edge joint, fabricated from 0.7 mm thick low carbon steel sheet, showing sheet thinning in one of the sheets caused by the fusion zone being slightly too wide and biased to one side of the joint.

component part (see Fig. 6.1). The larger the focusing f number the shallower the angle which can be achieved. For example, using f 7.5 the beam convergence angle will be approximately 7.5° whereas for f 3.5 the angle is just over 16°. Using the number f 7.5 the centreline of the laser beam can be angled at 7° from the plane of the joint line and have ample clearance, just over 3° to avoid clipping the protruding component part.

A laser beam impingement angle of 7 to 10° is recommended, if access to the joint permits, to ensure the weld penetration covers the maximum amount of the joint face. Fig. 6.14 (a) shows a cross section of a T butt joint where the laser beam was applied at an angle of 10° to the plane of the joint line and (b) a lap fillet joint where the beam was applied at approximately 30°. In the latter case barely 50% of the weld penetration has been usefully employed. An interesting feature of the T butt joint shown, apart from the more efficient use of the weld penetration, is that the fusion zone has curved and followed the joint line. Early workers thought the laser light 'skidded' down the joint line and this weld formation was often referred to as a 'skid weld'. The effect is actually caused by the inability of the T stem to conduct heat away as fast as the T crosspiece. This effect is of course an advantage, but it is lost when beam impingement angles greater than about 15° are used, see Fig. 2.4 in Chapter 2 for an example.

Laser beam/joint alignment for T butt and lap fillet joints requires special care and has a much smaller tolerance to variation than the focus position. Figure 6.15 shows the effect caused when a laser beam

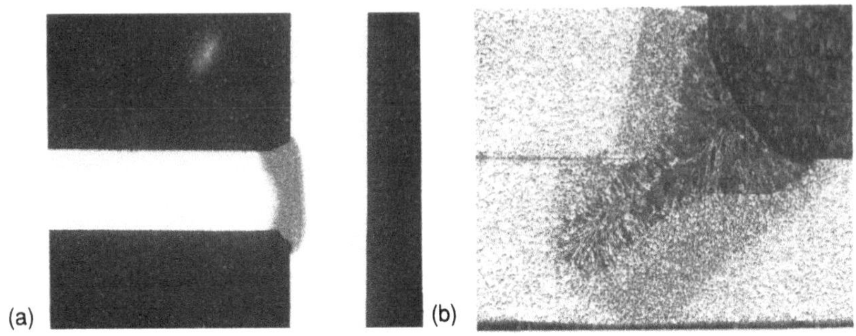


**6.14** Sections through a T butt joint (a) in 2 mm thick steel, where the laser beam impingement angle was 10° to the crosspiece and a lap fillet joint (b) in 1 mm thick steel, where the angle was approximately 30° to the sheet. The larger angle has caused much of the fusion zone to miss the joint line.

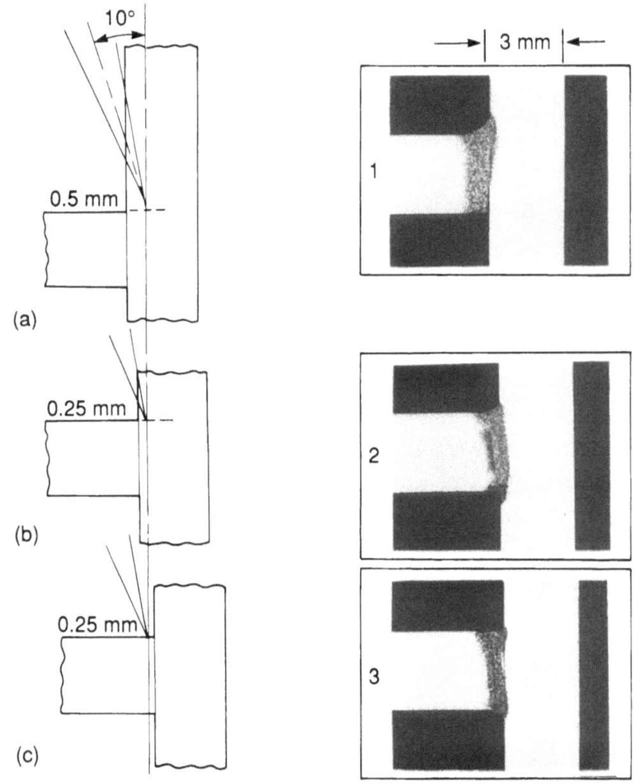


**6.15** Sections through low carbon steel test specimens used to establish the tolerance to laser beam/joint alignment when welding a T butt joint. Position (a) weld has only covered the joint at the top, (b) weld has encompassed the complete joint and (c) has only covered the lower portion. In all cases the focused beam minimum waist position was set at the surface of the T stem (focusing optic f number 7.5, calculated focus spot size 0.2 mm).

with a 10° impingement angle and a calculated minimum waist diameter of 0.2 mm is focused at the top of a T butt joint, in 3 mm thick steel, and then moved into the T crosspiece 0.25 and 0.5 mm, then moved away from the crosspiece 0.25 mm. At 0.5 mm into the crosspiece it can be seen that the laser beam has struck the crosspiece at a position above the joint and the shallow angle of incidence has caused the unfocused beam to reflect at an equal and opposite angle back on to the stem of the T and almost completely away from the joint line. At a distance of 0.25 mm into the crosspiece an acceptable joint coverage by the fused zone has occurred. However, at 0.25 mm away from the crosspiece, the fusion zone has failed to encompass the joint. In this instance the tolerance would be 0.1mm from the crosspiece and 0.25

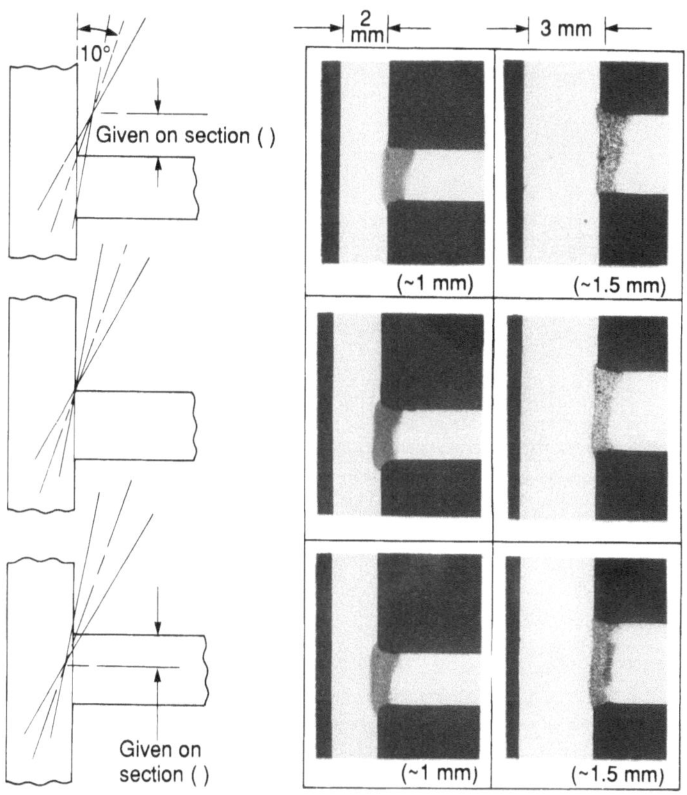


**6.16** Sections through low carbon steel test specimens showing the weld formation at the tolerances established for the focused beam minimum waist position about the work surface when making T butt joints in 2 and 3 mm thick material (focusing optic f number 7.5, calculated focus spot size 0.2 mm).

mm towards it. By comparison, at the zero position (the point where the surface of the stem and crosspiece which are exposed to the laser beam intersect) the laser beam minimum waist position (best focus position) could be displaced in or out ±1 mm for 2 mm material and ±1.5 mm for 3 mm material as shown by the test specimen presented in Fig. 6.16. The values presented should not be taken as a generalisation, as they will vary according to the overall welding parameters used.

Despite the poor tolerance to beam/joint alignment, the T butt and lap fillet joints have a wider tolerance to joint face mismatch than butt joints. This can be appreciated from the test samples shown in Fig. 6.17 for 2 mm thick joints in low carbon steels.

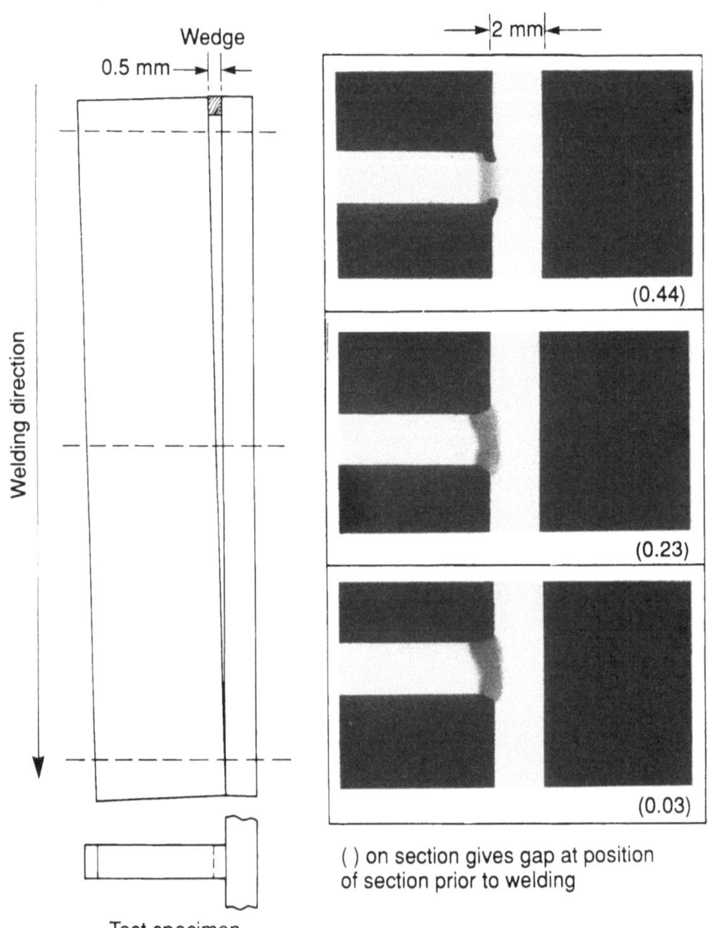


**6.17** Sections through low carbon steel test specimens showing welds produced in T butt joints with different joint face gaps.

## Laser beam spinning

Laser beam/joint alignment and joint fit-up require special care when welding butt joints in sheet metal and, as can be seen from this chapter, the tolerances to variation are small. Occasionally a situation arises where it is desirable to use laser welding but the required tolerances cannot be met, such as in roll formed tube butt joints, which form a narrow V between the joint faces, or long guillotined joints where straightness (and therefore fit-up) cannot be guaranteed. In such cases, a reduction in welding speed to increase weld width is ineffective as the small beam waist diameter will still clear the narrow and shallow joint gap. Increasing the focus beam minimum waist diameter, by increasing the f number is a possibility but usually impractical. To encompass a 0.25 mm gap between the faces of a butt joint and have sufficient power density incident on the opposing shoulders of the joint to effect a weld would require the focus spot, or minimum waist diameter depending on the term one wishes to use, to be doubled. This action is impractical because the laser power would have to be increased by a factor of four to maintain the required power density and thus necessitate using a much higher power laser than necessary for welding a relatively thin sheet material. Fortunately, this problem can be resolved without using a larger spot diameter, by using laser beam spinning, but at the expense of a reduction in welding speed of at least 20%.

The term beam spinning is not strictly correct since the focused beam axis is orbited about the position of its original axis prior to focusing, Fig. 6.18. This action together with its progression along the joint line (gap) between the workpieces produces a trajectory pattern, which is determined by the spin speed, orbit radius and workpiece velocity – the relationship is described below.

When welding poorly fitting joints in sheet metal a spin radius of only 0.5 mm can have a significant effect on weld formation. Figure 6.19 shows two roll formed steel tubes which were clinched together and had a 0.25 mm joint gap at the surface prior to laser welding. The weld made using beam spinning is visually wider and has a transverse cross section formation similar to that shown in Fig. 6.20, whereas the weld made without beam spinning is narrow and severely undercut.

Laser beam spinning can be achieved by numerous methods such as rotating a precisely tilted concave focusing mirror or conventional lens about its optical axis, or by rotating an angled optical flat below a transmissive or reflective focusing optic as shown in Fig. 6.21. The latter method can maintain the axis of the focused laser beam normal to the

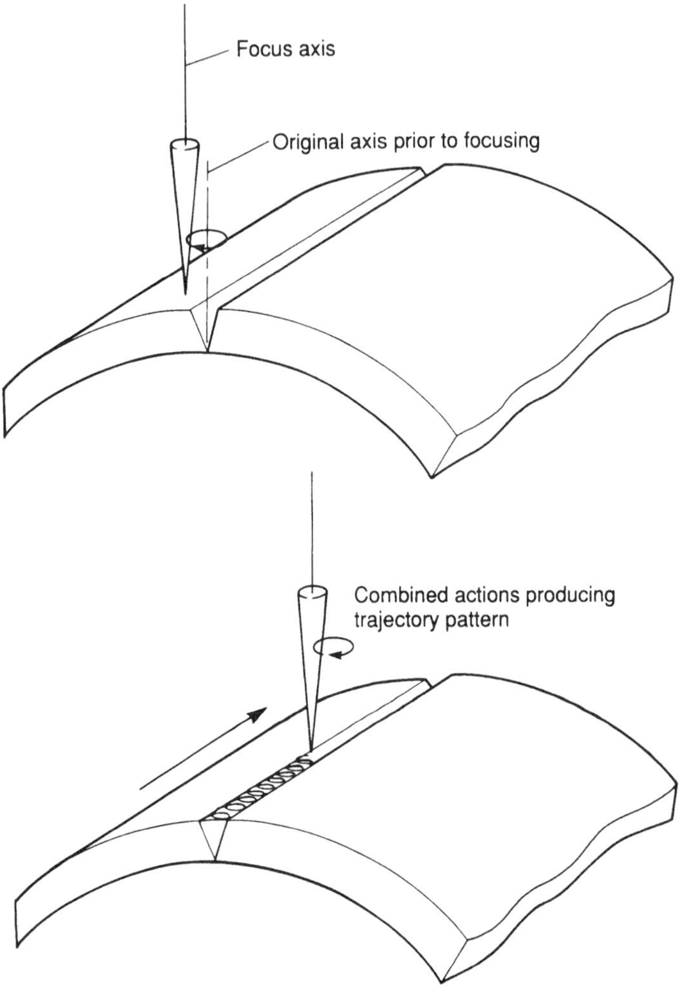


6.18 Principle of laser beam spinning operation.

work surface if required and is also much less sensitive to adjustment. However, transmissive optics are more delicate and prone to weld spatter damage. An apparatus using an optical flat, developed for use with a 5 kW, 40 mm diameter laser beam, is shown in Fig. 6.22. To achieve a spin radius of 0.5 mm a potassium chloride optical flat 6 mm thick at an angle of 15° can be used (the spin radius is determined by the

(a) (b)

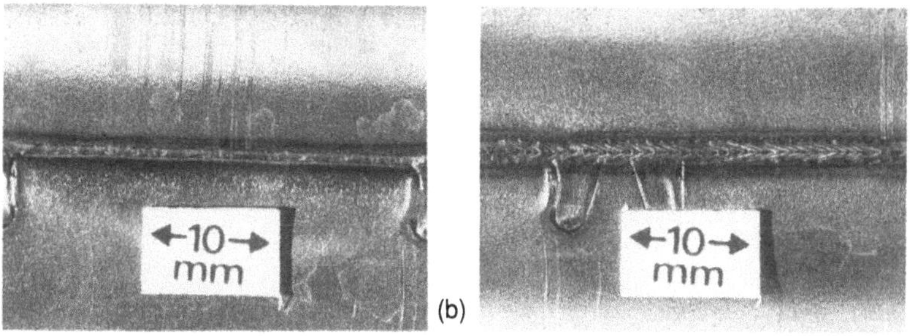


**6.19** Seam welds in roll formed steel tubes with post-weld gaps of up to 0.25 mm at the joint. The conventional laser weld (a) is narrow and badly undercut whereas the weld (b) made using laser beam spinning is wider and only slightly undercut.

(a)

(b)

(c)

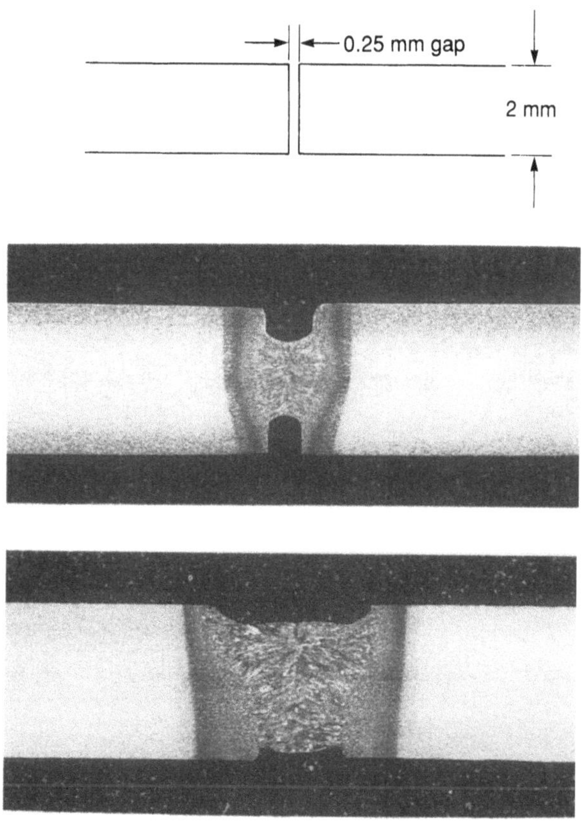


**6.20** Sections through conventional and beam spun laser welds: a) Butt joint test specimen set up prior to welding; b) Conventionally laser welded; c) Laser welded with beam spinning.

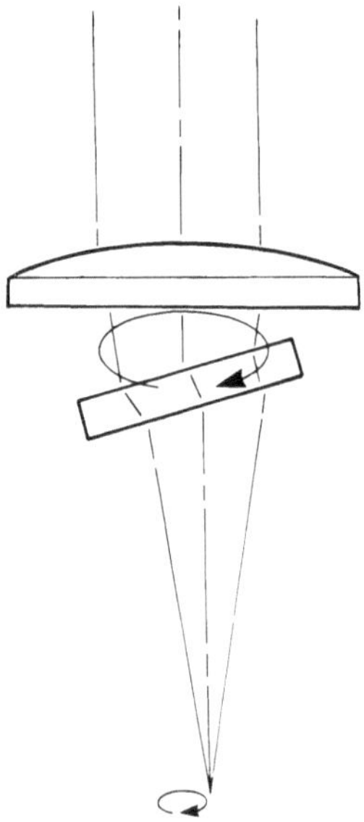

6.21 A method of achieving laser beam spinning by rotating an angled optical flat below a focusing optic. The spin radius is determined by the refractive index, thickness and tilt angle of the optical flat.

6.22 A laser beam spinning apparatus suitable for $CO_2$ lasers up to 5 kW in power. The integral belt guard and cover for the unfocused laser beam have been removed for clarity.

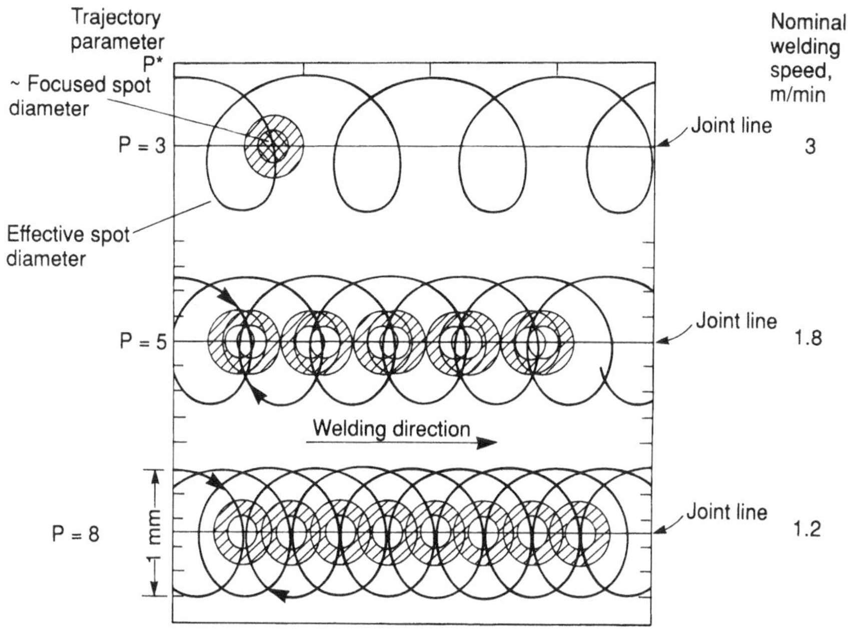


Traces of computer drawn beam spinning trajectories for: spin speed 3000 rev/min, (50 Hz) and 0.5 mm spin radius.

$$^{*}\text{Trajectory parameter } P = 2\,\pi \times \frac{\text{Spin speed (rev/sec) x spin radius, mm}}{\text{Workpiece velocity (welding speed, mm/sec)}}$$

**6.23** Examples of different heating patterns for different P numbers, where the effective spot diameter is equal to the spin radius.

refractive index, thickness and tilt angle of the optical flat). Note, potassium chloride should not be used where the humidity is high as this material is hygroscopic (absorbs moisture) and will dissolve, thus destroying its optical properties. See Chapter 11 for details of optical materials.

When using beam spinning the correct selection of trajectory pattern is most important, otherwise a cyclic or intermittent weld formation occurs. The selection is not as difficult as it may at first sound if the following formula is applied:

$$P = \frac{2\Pi fr}{w},$$

where P = the trajectory pattern number.
f = rotational frequency, rev/sec.
r = spin radius, mm.
w = workpiece velocity, mm/sec.

To avoid cyclic or intermittent weld formations there are two important rules which must be observed:

1) P should be approximately 8 or more;
2) The spin radius should not exceed the effective spot diameter if the value of P cannot be increased beyond 8 by controlling the spin and welding speeds.

To help understand all this, Fig. 6.23 shows tracings from computer drawn trajectories with the focus spot and effective spot diameters superimposed on the joint line. Where P = 3 an intermittent joint would be formed and where P = 5 an uneven cyclic heating and welding pattern would occur, whereas at P = 8 the weld would be continuous and its formation more uniform.

Chapter 7

# Welding plate, tube and pre-machined parts

Research by the United States Navy, in the mid-1970s, using a 15 kW $CO_2$ laser, proved single pass weld depths of 18 mm in steel could be achieved. Subsequently their contractors, during the 1980s, produced laser welded missile launch rails and aircraft carrier catapult rails, each several metres long, with weld depths in the order of 18 mm.[1] Despite this advanced work, which had to meet stringent service requirements, plus numerous other successful pilot studies for welding 12 to 18 mm thick material conducted in the United Kingdom, Europe and Japan, most commercial applications, where plate and tube or indeed tube to plate joints are made by laser welding, the weld depths are generally only between 3 and 6 mm. The delay in exploiting the technology for welding thicker sections is usually attributed to the cost of the necessary high power laser, the associated sophisticated laser beam or work manipulation equipment and the high cost of material preparation and positioning; it is costly to machine long square edges for close fitting joints and heavy components are difficult to position and clamp accurately. Nevertheless, as the new design and volume production opportunities offered by laser welding are grasped, suitable applications will no doubt come to light which justify the high capital and processing costs, such as the manufacture of special section engineering beams, large engine crank cases and precision machine tools and processing plant. With these eventual applications in mind this chapter considers information and data pertinent to welding with laser powers up to 10 kW and weld depths up to 12.5 mm.

## Welding conditions

Although a method for establishing welding conditions was described in Chapter 5 the subject is briefly returned to here, because as laser weld penetration depth increases beyond 6 mm weld quality can be more variable and thus acceptable weldability envelopes become smaller. This is because as laser weld depths increase the aspect ratio of weld depth to keyhole diameter also increases. The resultant deep and narrow keyholes restrict outgassing and if the welding conditions are not carefully defined entrapped gas causes porous welds. In steels, the gas formation can be the result of a high oxygen content in the material (see Chapter 4 for comments on oxygen levels in steels) or the decomposition of oxides (e.g. inclusions) and possibly the entrapment of shielding gas. To minimise weld porosity a careful study of laser power and welding speed in relation to the focusing optic and gas shielding conditions is essential. An example of a weldability envelope from such a study, together with sections showing the variations in weld profiles and porosity formation in relation to different welding conditions, is shown in Fig. 7.1. The study, by Watson,[2] was made for welding a 12.5 mm thick structural steel, type BS 4360 Grade 50D, and shows there are two types of porosity formation:

1) Large pores in the centre of the lower half of the fusion zone thought to be due to poor keyhole formation at the weld root caused by using a welding speed close to the upper limit for achieving the full 12.5 mm deep penetration at the laser power used;
2) Fine, more evenly distributed pores, usually found near the fusion boundary and caused by low welding speeds which allow the gas to nucleate and grow into small bubbles as the metal is molten for relatively long times.

Before leaving the subject of welding conditions it is worth reminding oneself that plasma control (see Chapter 5) is necessary when making autogenous welds where the laser power available limits the welding speed to below 1 m/min. Chapter 5 describes the necessary plasma control equipment and gives a guide to its set up, the plasma control gas and gas flow rate.

## Joint configurations

Many of the joint configurations already shown in Chapters 2 and 6 also apply to this chapter. However, joints in thick material sections can be fabricated by making welds from each side of the joint as shown by the weld cross sections in Fig. 7.2. There are other joint configurations which have not been previously mentioned and which can

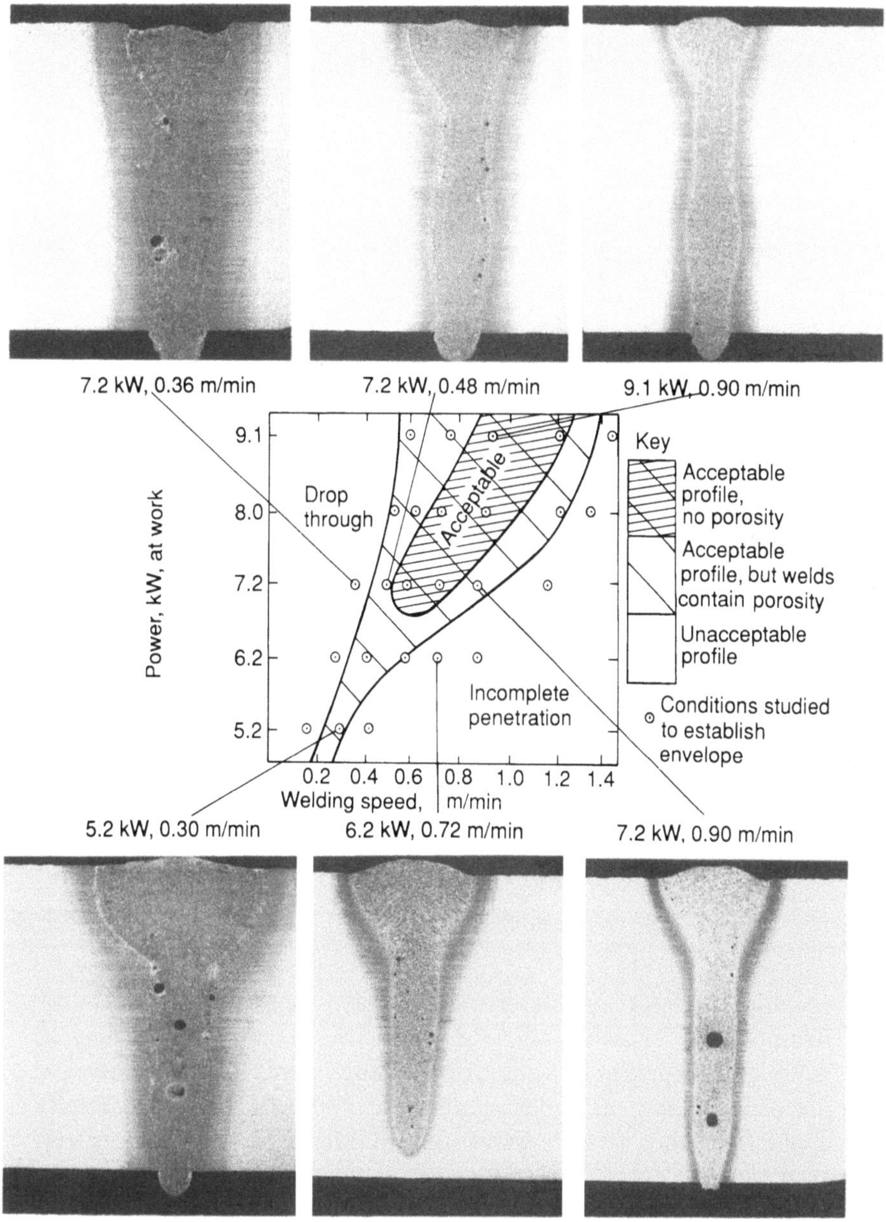


**7.1** Weldability envelope and transverse cross sections of weld profiles found when establishing conditions for $CO_2$ laser welding 12.5 mm thick butt joints in BS 4360, Grade 50D, steel.[2]

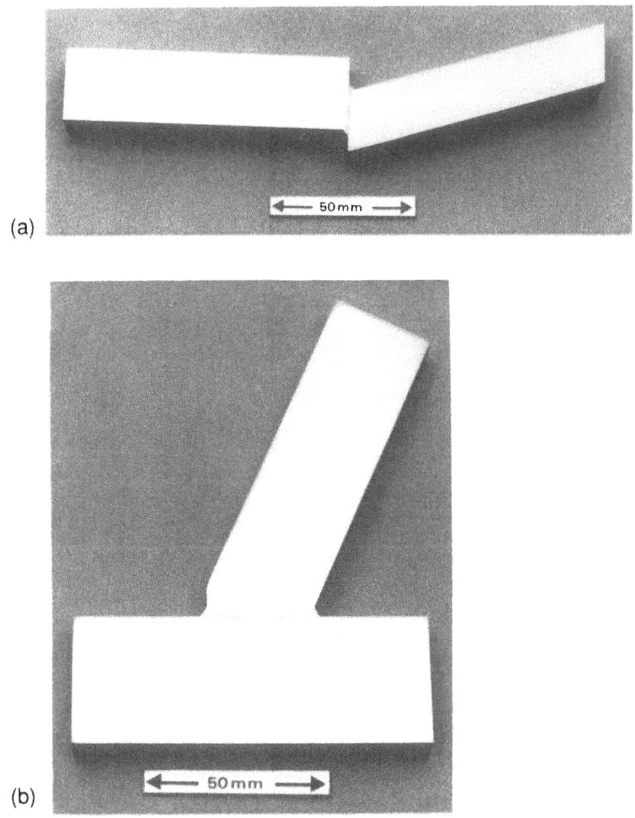


**7.2** Examples of double sided $CO_2$ laser welds made in stainless steel materials: a) Two abutting weld penetrations in a staggered butt joint; b) Two partial penetrations forming a T butt joint.

be called upon to assist laser beam to joint access and part location. Examples are shown in Fig. 7.3 which are practical to engineer in material thicknesses of 3 mm or more. Such joints lend themselves to precision equipment and machine tool fabrication and if carefully applied, together with the low distortion offered by laser welding, can keep post-weld machining allowance to a minimum.

When fabricating intricate structures, such as multi-box sections, the different joint configurations can be used collectively and to advantage. In such cases, the component part clamping and welding sequences need to be carefully considered at the product design stage to allow for weld shrinkage and distortion, which although quite small for an individual weld, can collectively affect joint fit-up in the later

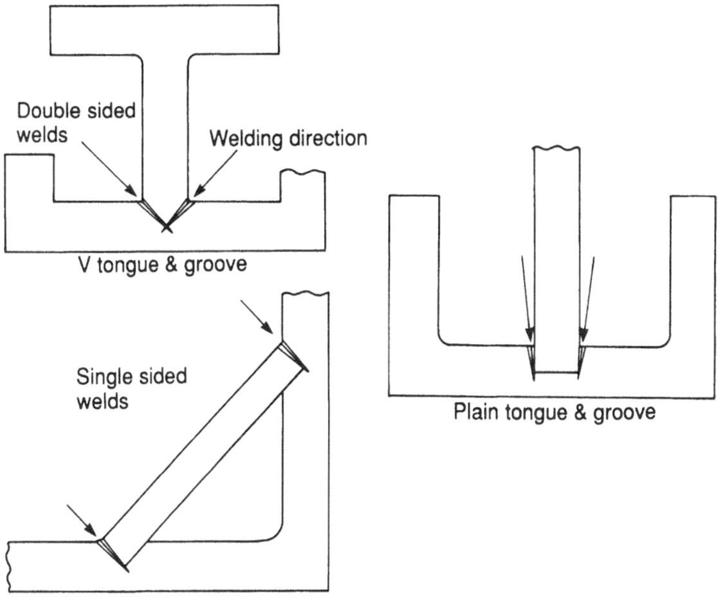


**7.3** Cross sections of joint configurations which can assist laser beam to joint access and part location.

stages of the welding sequence. Figure 7.4 shows a hypothetical component which could be laser welded and the sequence of operations. The numbers depict the order of welding and the broad arrows the direction(s) of clamping during each welding operation. In the example shown, starting from the centre and leaving the corner edge butt joints until last avoids ending up with unwanted gaps between the plain butt joints covered by welding sequence numbers 5 and 6.

## Joint fit-up and laser beam/joint alignment

The necessary considerations for joint fit-up and laser beam alignment are the same as for sheet steel joints, described in Chapter 6. The actual values have to be established according to the dimensions of the transverse cross section of the weld profile produced by the welding conditions used and in relation to the chosen joint configuration.

Generally, as material thickness increases the tolerance to gaps at the joint face also increases and the effect on weld undercutting of the plate surface or weld slumping, becomes less pronounced than in thin

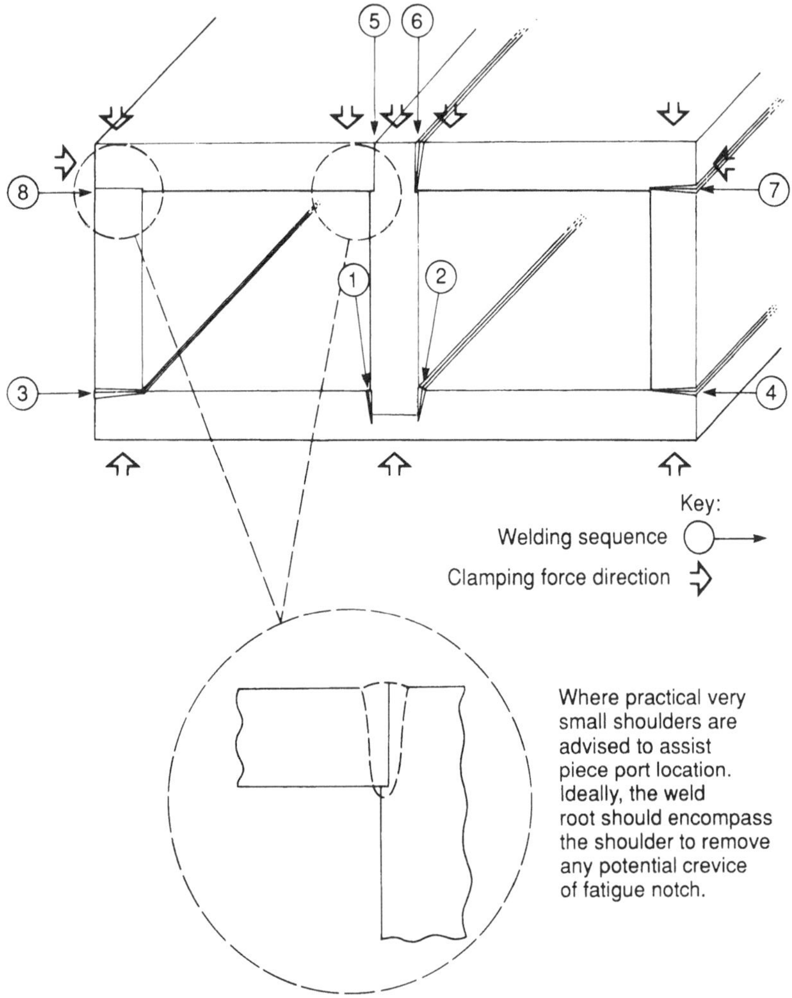


**7.4** A hypothetical example of a twin box section fabrication employing two different joint configurations. The welding sequence and required clamping force directions are also shown.

sheet, due to the increased aspect ratio of weld depth to width. Figure 7.5 shows cross sections of autogenous welds made in 8 mm thick steel plates, which prior to welding had gaps between their butt joint faces of 0.25 and 0.75 mm. It should be noted in these examples that plasma control (described in Chapter 5) was deliberately not used and the welding speed kept low to widen the weld fusion zone to encompass

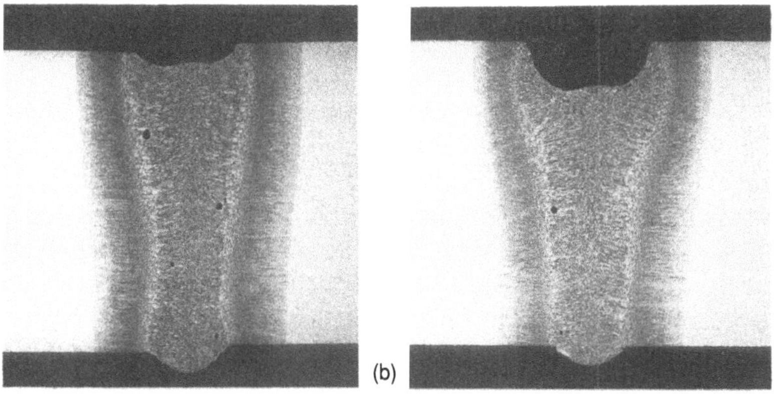

(a) (b)

**7.5** Transverse cross sections of weld profiles obtained in butt joints between 8 mm thick steel plates with gaps at the joint face before welding of 0.25 mm (a) and 0.75 mm (b). Note the increase in undercutting in relation to the pre-weld gap.

the shoulders of the joints. Furthermore, the wider gap (0.75 mm) required a much lower welding speed to achieve this objective; 0.35 m/min compared with 0.5 m/min. The level of weld undercut shown in Fig. 7.5(b) may be unacceptable for certain applications, necessitating the use of a filler wire as described in Chapter 8.

The tolerance for sideways alignment of the laser beam with the joint line is also increased with increased plate thickness, but not by a substantial amount. This is despite the higher laser power, larger associated focus spot size and reduced welding speed necessary for welding thicker sections. For example, a weld made in 12.5 mm thick steel at a laser power of 9 kW, welding speed of 0.9 m/min and a focus f number of 7.5, will have a tolerance of little more than 0.7 mm each side of the joint line. It is easy to be tempted to estimate the tolerance by measuring the external width of the weld underbead. This approach is not recommended as welds in some materials, especially stainless steel can be narrower in the material through thickness; refer to Fig. 7.5(a) for an example in low carbon steel.

## Joint tacking

Joint fit-up and beam alignment are extremely important but their careful set up can be lost when welding thick sections, especially where long weld lengths are required, if adequate joint clamping (see Chapter 10) and pre-weld tacking are not employed. Tacking is very important since weld shrinkage can apply considerable stress within the work and open up a joint despite employing substantial compressive clamping forces applied to the joint.

Single spot tacks are not recommended as they have poor strength. The best results are achieved by using tack welds with a weld depth equal to that ultimately required and with a length equal to twice the weld depth, e.g., a full penetration weld in a butt joint between 12.5 mm thick plates would have a tack weld length of 25 mm. Generally, tack welds need to be placed on a 300 mm pitch on long lengths of material, especially where the plate rolling direction and hence stress relief pattern is unknown. The first tack should be made in the centre of the required weld length and the other tackings should be made working away from each side of centre. This approach helps to maintain dimensional accuracy and keeps post-weld straightening of long components such as I beams to a minimum.

Welds made over tack welds often do not have smooth profiles and tend to contain more porosity than the weld metal between tacks. During the welding operation the welding sound also changes as the tacks are crossed. These effects are thought to be due to welding a recast material structure and the absence of abutting joint faces influencing outgassing.

## Closing the keyhole of circumferential welds

When making circumferential welds which have to overlap their start position, such as when welding tube to plate and tubular butt joints, two problems can occur if the laser welding power is suddenly switched off. Firstly, a deep depression is left on the work surface where the molten walls of the final keyhole position slump together under the influence of gravity and capillary attraction. Secondly, as the keyhole walls draw together, the high weld contraction forces can cause serious cracking at the point of final closure. Moreover, both problems are exacerbated as the material thickness to be welded is increased. These problems can be considerably reduced and sometimes, depending on the materials, can be eliminated by ramping down the laser power and/or accelerating the welding speed. The satisfactory effect of these actions, however, requires careful trial and error.

Ramping down the laser power has to be carried out quickly if a large weld overlap and an unnecessarily long weld cycle time is to be avoided. The gradual reduction in weld penetration when ramping can produce a slanting string of porosity along the ramped weld length, due to the blind keyhole formation and welding over a previous weld. Generally, in weldable grades of steel, ramping can achieve acceptable results in weld depths up to 5–6 mm, however, it can be difficult to eliminate the root porosity and cracking in thicker sections.

Ramping by increasing the welding speed alone, and then switching off the laser power can be achieved if the initial welding speed is in the region of 1 m/min, or less, and a substantial increase in welding speed of several metres per minute, with rapid acceleration is used. Defocusing the laser beam in conjunction with increasing the welding speed is another possibility, but very few $CO_2$ lasers are equipped with automatic focus control. Moreover, accurately resetting focus position without a precision control system can present repeatability problems in itself.

## Welding partial penetration joints

In Chapter 4 materials prone to solidification and post-weld cracking, e.g. steels with high carbon and/or sulphur and phosphorus levels, were discussed. One may recall that the high weld contraction stresses, as the weld shrinks on cooling, promote and initiate cracking. In partial penetration welds, crack initiation is exacerbated by the fact that the solid abutting metal at the weld root opposes weld shrinkage and promotes even higher contraction stresses than a full penetration weld. However, in many cases this problem can be overcome by employing a weld stress relieving groove.

The arrangement of a stress relieving groove is shown in Fig. 7.6 and such grooves are often employed when making fusion welds in tube to plate or flange and gear to shaft assemblies. The groove, which runs parallel to the joint line is deeper than the required weld penetration depth and has a radiused root to avoid any sharp notch which could lead to a source for in-service fatigue failure. The groove provides one side of the joint with a narrow flange, in the case of a conventional circumferential butt joint, or a narrow sleeve in the case of an annular joint. In both cases the groove breaks the heat conduction path into the bulk material and this significantly reduces the rate of weld cooling. Consequently, during the welding operation the flange, or sleeve as the case may be, is heated more than normal, thus its mechanical strength is reduced, and it is able to flex and be drawn towards the weld during the weld solidification period, reducing weld contraction stress considerably and eliminating weld cracking in many instances.

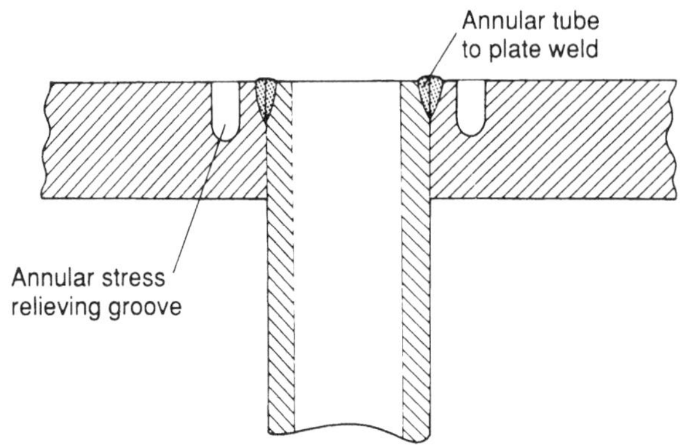


**7.6** Schematic arrangement of an annular stress relieving groove in relation to a tube to plate joint.

Unfortunately, there is not an established standard for the stress relieving groove dimensions in relation to laser weld depth and therefore some experimentation is necessary to ensure the groove wall will flex sufficiently and also have adequate mechanical strength to meet its service requirement. The author has successfully used stress relieving grooves with a root depth equal to twice the weld depth and a wall thickness nominally equal to the weld depth, see Fig. 7.7. However, the thinner the wall which can be tolerated, the lower the weld stress. Furthermore, the welding speed should be kept low to get maximum heat into the wall.

Partial penetration welds which do not suffer from solidification cracking, or where cracking is eliminated by a stress relieving groove, can be prone to cold and/or fatigue cracking initiated by the natural notch crack provided by the unwelded joint faces, and which runs right up against the weld root. Where partial penetration welds are made from both sides of a component, as in Fig. 7.2(b), the likelihood of these types of cracking is much lower than for a single sided partial penetration welded joint. In the latter case continuous or cyclic loading can cause the notch, which provides a sharp notch at the weld root, to propagate into and tear the weld, see Fig. 7.8 for a typical example. A method of eliminating this effect is to machine a notch blunting groove into one or both of the joint faces; this has a similar effect to arresting a tear in a sheet by drilling a hole at the end of the tear.

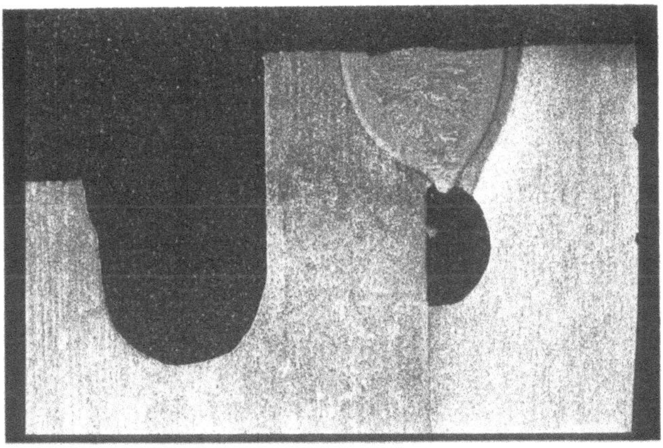

7.7 A transverse cross section through a 4 mm deep weld in an annular butt joint employing a stress relieving groove and a semi-circular notch blunting groove. The materials are 0.2 carbon steel and the weld was made at 4.5 kW and 1.2 m/min.

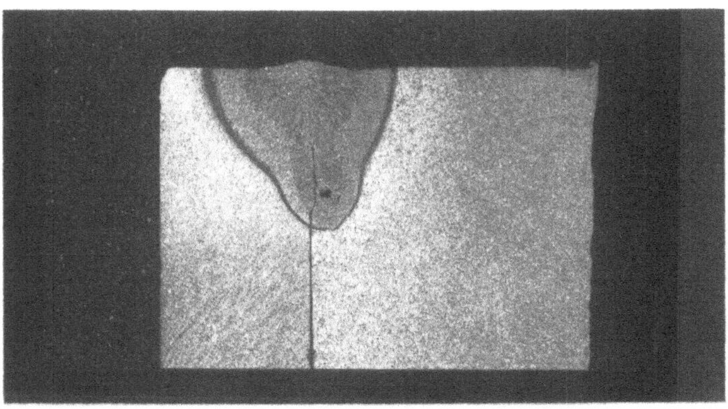

7.8 An example of a crack which has propagated from the sharp notch, at the weld root, produced by the unwelded joint faces of a partial penetration weld.

Notch blunting grooves should ideally be symmetric about the weld root, however, it is not essential. Reference to Fig. 7.7 shows where a notch blunting groove is only used in one side of the joint, as it is not possible to use it in both sides when used in conjunction with a stress relieving groove. The radius of the notch blunting groove needs to be considered in relation to size of the weld underbead width. Ideally the radius should be such that it provides the maximum notch angle at the weld root and also is practical for machining.

## References

1 Irvine R R: 'The high powered laser welds carbon steel plate'. Ion Age 1982 11(8) 48–49.
2 Watson M N: 'Laser welding of a 12.5 mm thick niobium microalloyed structural steel at powers of up to 9 kW'. Weld Inst Res Report 243/1984.

Chapter 8

# Single and multi-pass welding with filler materials

Laser welding with filler materials is still very much in its infancy in terms of development and exploitation and therefore, at present, there are few time-proven industrial applications from which to draw experience. Perhaps one of the first, and most proven production applications was developed by Kawasaki Steel Corporation in Japan, where filler wire is used when laser welding butt joints in silicon steel for continuous strip production.[1] Although at present limited, the use of filler materials in conjunction with laser welding will no doubt broaden as the advantages of this relatively new technology are grasped and production welded components and structures are designed with laser welding in mind.

There are three main reasons for using a filler material:

1 To improve weld properties;
2 To fill gaps where acceptable joint fit-up for autogenous welding cannot be achieved;
3 To achieve multi-pass welds and enable the laser to weld sections several times thicker than its normal single pass, autogenous welding capability.

Filler material can be applied to the weld region in several ways; via a continuous wire feed system; in the form of a pre-placed wire; or as a thin strip nipped between the joint faces. Other techniques such as powder addition are being researched.

This chapter first briefly discusses filler materials and then presents information on applying filler metal in the continuous wire feed and pre-placed forms. It also describes joint profiles, which have been successfully used to make multi-pass welds.

## Filler materials

At present there are no commercially available filler wires which have been specifically developed for laser welding. Consequently, metal inert gas (MIG) welding wires are usually used where filler wire is applied in the continuous wire feed form. Filler materials are designed according to the materials to be joined and the welding technique used; steel alloy MIG wires, for example, are copper-coated to assist the application of the arc current and some elements in the wire are carefully balanced to maintain arc stability and metal transfer. The weld formation and heating regime of a laser weld is quite different to MIG and other arc welding techniques (see Chapter 2) and therefore it cannot be assumed that all MIG wires are ideal for use with laser welding. More wire and parent material research must be completed before well established filler wire data can be presented for laser welding. Nevertheless, the major wire and parent metal manufacturers will usually make helpful suggestions with respect to using MIG welding wires and should always be approached for initial guidance if necessary.

Experience when laser welding C-Mn-Si-Al-Nb structural steels, such as BS 4360, grade D, has shown that MIG welding filler wire to BS 2901, Part 1, 1970, types A15, A18 and A19 improves weld toughness and opposes solidification cracking. Likewise, good results have been achieved when welding stainless steel type 304L using BS 2901.2 308 S 93 stainless steel wire. The basic compositions of the filler wires described are presented in Table 8.1. In the case of the steel wires, similar wire materials, but with lower carbon levels, may be better, since low hardenability is necessary to give good weld impact toughness properties; one may recall from Chapter 4, that the high cooling rate of a keyhole laser weld causes a much higher weld hardness than the slower cooling rate of an arc weld. Nevertheless, laser welds made with filler wire are generally much wider than autogenous laser welds, they are usually made at slower speeds and in such circumstances can have much reduced weld hardness levels.

## Welding with continuous wire feed

Before proceeding further, those considering using laser welding with continuously fed filler wire, to make welds where high quality is required, should be aware that the amount of side wall porosity experienced in single pass laser welds, tends to be higher than when making autogenous welds, and also tends to increase with joint gap size. Weld root porosity is also often produced when making multi-pass laser welds. Moreover, there appears to be a complex relationship between porosity and filler wire type, wire diameter, joint gap, gas shielding

*Table 8.1* **Basic chemical analysis, wt%, of MIG welding filler wires successfully used when laser welding C-Mn-Si-Al-Nb structural steel and type 304L stainless steel**

| C-Mn-Si-Al-Nb steel Wire type | C | S | P | Si | Mn | Ni | Cr | Mo | Cu |
|---|---|---|---|---|---|---|---|---|---|
| BS 2901 Ti-treated A15 | 0.11 | 0.015 | 0.011 | 0.79 | 1.31 | 0.06 | 0.06 | 0.01 | 0.45 |
| Double deoxidised A18 | 0.07 | 0.012 | 0.012 | 0.91 | 1.40 | 0.02 | 0.01 | 0.01 | 0.36 |
| Al-treated A19 | 0.12 | 0.014 | 0.014 | 0.47 | 1.25 | 0.03 | 0.01 | 0.01 | 0.18 |
| **304L stainless steel Wire type** | **C** | **S** | **P** | **Si** | **Mn** | **Ni** | **Cr** | **Mo** | **Cu** |
| BS 2901.2 308 S 93 | 0.009 | 0.008 | 0.019 | 0.84 | 1.69 | 10.34 | 20.02 | 0.06 | 0.05 |

conditions, welding power and speed. Consequently, some careful experimentation and weld examination for a given application is usually necessary. Research work is currently in hand to improve filler wires and procedures to reduce porosity and improve weld properties.

Feeding the filler wire into a parallel joint gap which is slightly wider than the diameter of the filler wire is the recommended practice whenever possible. Adding filler wire to joint gaps which are narrower than the wire diameter is difficult to achieve, without special equipment and precise control, because in such circumstances the wire has to be placed right on the edge of the keyhole to heat consistently, melt and feed into the weld zone and to achieve these requirements the filler wire positioning tolerance, about the edge of the keyhole, is extremely small. Without precise positioning and control, which could be difficult to achieve in a heavy engineering workshop, the filler wire has only to move little more than a millimetre about the keyhole to either a) project into the beam path and cause excessive spatter, vaporise and form a plasma cloud which defocuses the laser beam and re-radiate its energy elsewhere, or b) move back from, or to the side of the keyhole, causing the wire to either stick to the work surface and possibly jam the wire feed unit, or alternatively, cause the unheated wire to snake off in another direction. The recommended approach of feeding the filler wire into a gap slightly larger than the wire diameter, considerably reduces the risk of these problems because it increases the wire positional tolerance in relation to the weld zone to a practical working value. Therefore, this approach will be described using a laser power of 5 kW and welding 8 mm thick low carbon and stainless steels as a model. The description covers the necessary equipment and provides set-up data. Welding of 8 mm thick material has been chosen as it is considered to be the maximum practical thickness for single pass welding with a 5 kW laser when using filler wire. It is hoped that from the information presented and reasons given, the reader will be able to extrapolate sufficient guidance for successful experimentation with other laser powers and material thicknesses.

When butt welding 8 mm thick plates it is possible to fill gaps up to 2 mm wide without using a backing bar. However, from a practical standpoint, a 1.5 mm gap and a 1.2 mm filler wire diameter is a good combination for ease of operation; the wire is relatively stiff and fits and moves nicely in the gap. Furthermore, when such a combination is welded using a laser power of 5 kW and a focus number of f6, a welding speed, for stainless steel, can be achieved in the region of 0.6 m/min (a slightly lower speed is necessary for welding low carbon steel).

The wire feed rate for a given gap and plate thickness is an important parameter and is dependent on welding speed, the cross sectional area of the gap between the joint faces and the cross sectional area of the filler wire. The relationship is expressed as follows:

$$\text{Wire feed rate, m/min} = \frac{\text{Welding speed, m/min} \times \text{Cross sectional area of gap, mm}^2}{\text{Cross sectional area of filler wire, mm}^2}$$

Using a 1.2 mm diameter filler wire to fill a gap of 1.5 mm between the faces of a butt joint between two 8 mm thick plates, to be welded at 0.6 m/min, the above formula establishes that wire feed equipment capable of delivering wire at a rate of approximately 6.4 m/min is required. (It should be noted here that small changes in the joint gap require changes in wire speed to maintain uniform gap filling. Consequently, for applications where variable gaps are envisaged, some form of gap sensing system would be required to detect changes in gap cross sectional area and adjust the wire feed to suit.)

Filler wire fed directly off the supply spool is pre-curved and has a helix angle, formed when the wire was coiled on the spool. In this condition the wire cannot be pointed in a straight line, which is essential to maintain the accurate pointing stability necessary for positioning the tip of the wire in relation to the laser beam and weld pool. To meet this requirement it is necessary to have a wire feed unit, Fig. 8.1, fitted

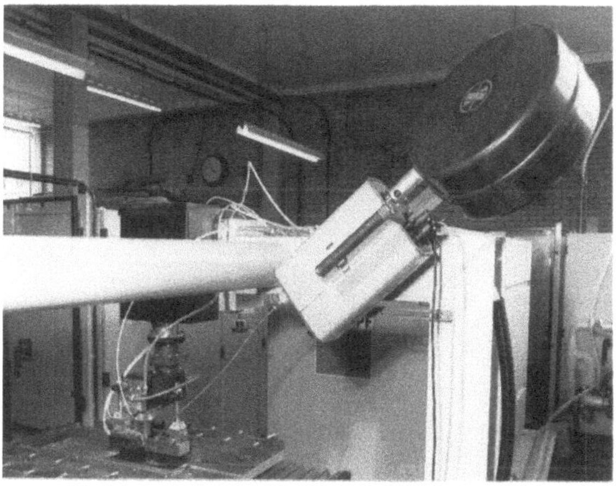

**8.1** A wire feed unit and supply spool, foreground. This unit has an integral planetary wire straightener. Leading from the unit, at approximately 45° is the sheath in which the filler wire is fed to the wire feed shoe, lower left.

with an auxiliary planetary wire straightening device; planetary devices are effective and deliver an almost needle-straight wire.

A normal tungsten inert gas (TIG) wire feed nozzle and positioner mechanism can be used to place the filler wire in the weld zone. These apparatus, however, tend to be delicate and are easily disturbed and therefore a more robust wire feed shoe device with integral X, Y and Z positional controls is recommended. Such a device is shown in Fig. 8.2 and is based on the plasma suppression device detailed in Chapter 5, Fig. 5.15. The major difference is that the shoe length needs to be increased to 130 mm to allow the sheath, delivering the filler wire, to pass the vertical height adjustment column more easily. Also when feeding the filler wire in the weld pool leading edge, the shoe gas shield is unnecessary. Another difference is the nozzle; a design which has been successfully used is shown in Fig. 8.3.

The successful operation of feeding filler wire depends on the wire feed delivery angle and position in relation to the laser beam and weld zone. Angles between 30 and 60° from the vertical can be used and 45° tends to be the norm, as it simplifies setting the required wire intersection position with the laser beam centreline, as will be seen shortly. Angles greater than 60° make the latter difficult and angles less than 30° cause the wire to intersect a large area of the laser beam, causing melting and vaporisation of the wire without incorporating it in the weld pool.

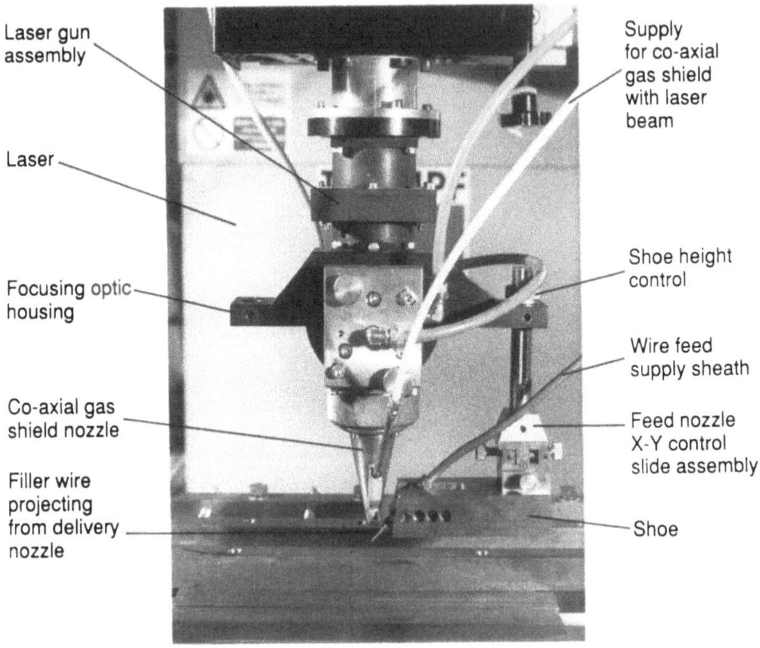


**8.2** A wire feed shoe suspended from a laser gun assembly.

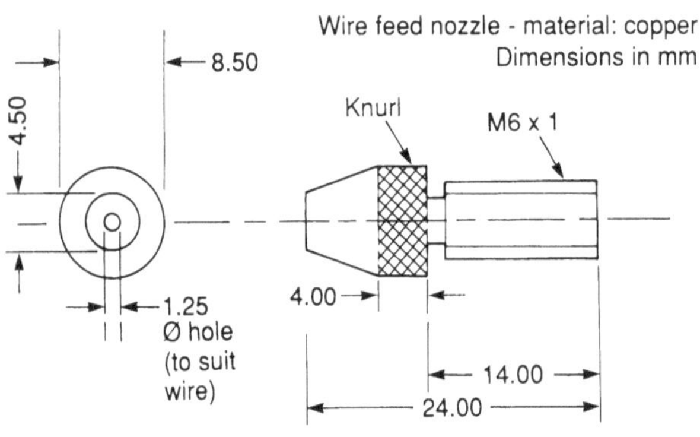


**8.3** Details of a wire feed nozzle which can be used in conjunction with a wire feed shoe developed from the plasma suppression device detailed in Chapter 5, Fig. 5.15.

The filler wire can be fed into the leading or trailing edge of the molten weld pool. The normal practice is to feed into the front, because when feeding into the trailing edge incomplete mixing of the wire into the pool occurs, see Fig. 8.4. This may not matter if the purpose of the filler wire is solely to fill a gap, but can be undesirable if the wire addition is intended to modify the weld metal properties in some way.

The position where the filler wire meets the weld pool is perhaps the most important setting for the smooth transfer of feed wire into the pool. The wire must not intersect the laser beam above the weld pool as this will cause excessive spatter, tooling damage and reduced penetration. When welding 8 mm thick plates the best position for intersecting the laser beam centreline is 3–3.5 mm below the plate surface, as shown schematically in Fig. 8.5; in reality the weld pool extends forward of this position and the wire strikes it nearer the plate surface. Setting the beam centreline/wire intersection position is made easy with the filler wire feed at 45° to the plate by the fact that if the wire is positioned to strike the plate surface at a point 3–3.5 mm from where the laser beam meets the work, the wire will also impinge on the beam centreline at the same distance below the work surface. The beam focus position for the welding example given is best set at the work surface, although this is not critical as no significant difference will be observed if it is set 1.5 mm above or below the surface.

Gas shielding of the weld pool is important when welding with filler wire, especially where a gap is present between the plates to be joined, and is more important than when making autogenous welds, because the weld pool is much larger and longer than the autogenous laser weld keyhole. The relatively long trailing edge of the weld is particularly vulnerable and a trailing gas shield is necessary. When feeding the filler wire into the leading edge of the weld pool the shoe device (see Fig. 8.2) cannot provide the trailing gas shield facility. However, the plasma suppression shoe device (Chapter 5, Fig. 5.15) can be used in tandem, but with the plasma suppression jet removed (a plasma is wanted in the joint gap to re-radiate the beam energy on to the joint sidewalls and cause the melting which is necessary to start the process). Using the tandem shoe arrangement, Fig. 8.6, for our 8 mm thick plate model, the following helium shielding gas flow rates are suggested: 15 l/min through the gas shield nozzle co-axial with the laser beam; 20 l/min through the plasma suppression jet holder (jet removed); 20 l/min through the shoe which provides the trailing shield.

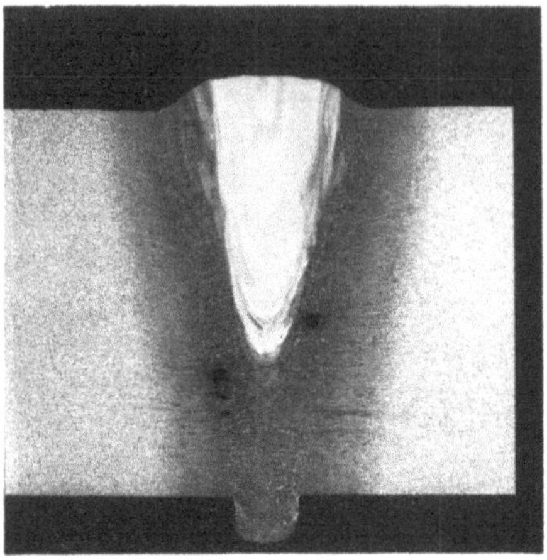

**8.4** Transverse cross section of a weld made with the filler wire fed into the trailing edge of the weld pool. Note incomplete mixing.

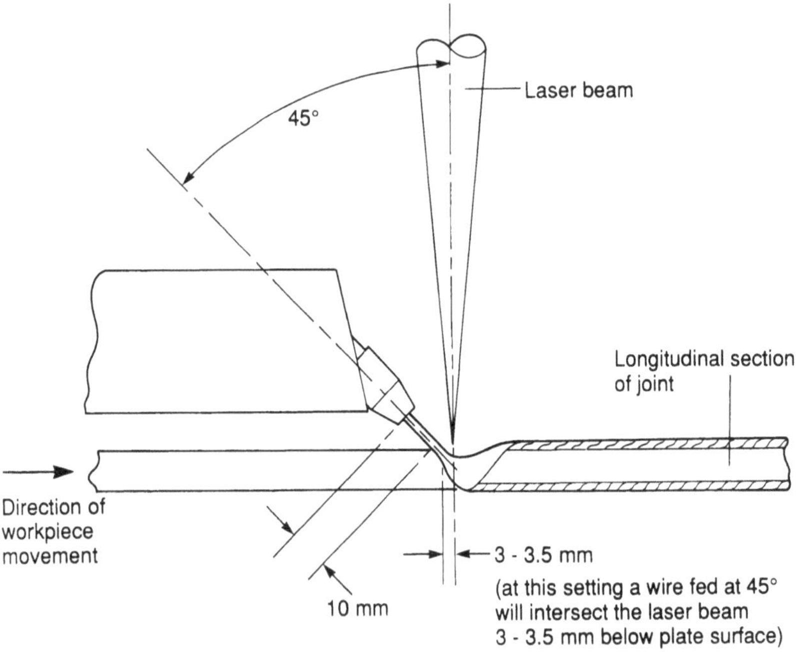


**8.5** Geometry of wire feed set-up for feeding into the leading edge of the weld pool when welding 8 mm thick plates, with a 1.5 mm gap between the joint faces and using a 1.2 mm diameter filler wire.

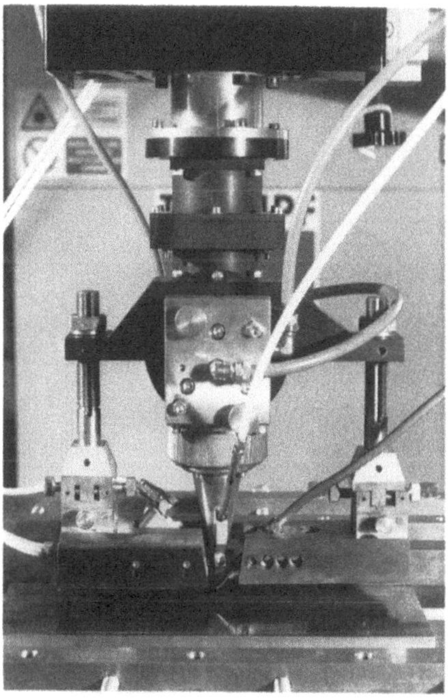

8.6 A plasma suppression device, left, used in tandem with a wire feed shoe, right, to provide a trailing gas shield when the filler wire is fed into the weld pool leading edge.

The information presented above for butt welding 8 mm thick steel and stainless steel plates using filler wire and a laser power of 5 kW will enable one to achieve quick success and gain some first hand practical experience. For thinner plates, the filler wire/laser beam centreline intersection distance will be less, but slightly greater for thicker plates. However, for plates thicker than 8 mm a laser power more than 5 kW is required to achieve a single pass weld. Nevertheless, from the model used it can be appreciated that a 5 kW laser can multi-pass weld material thicknesses greater than 8 mm by combining autogenous and wire fed welding operations. To illustrate this a section through a four pass weld made in 25 mm thick steel, using a laser power of 5 kW, is shown in Fig. 8.7. No such industrial applications are known to the author, but the technique has potential industrial possibilities and deserves further development.

The approach to laser welding with filler wire described above, refers to welding with cold wire. It is possible to weld with hot wire, that is a wire which is electrically resistance heated just prior to entering the

(a)

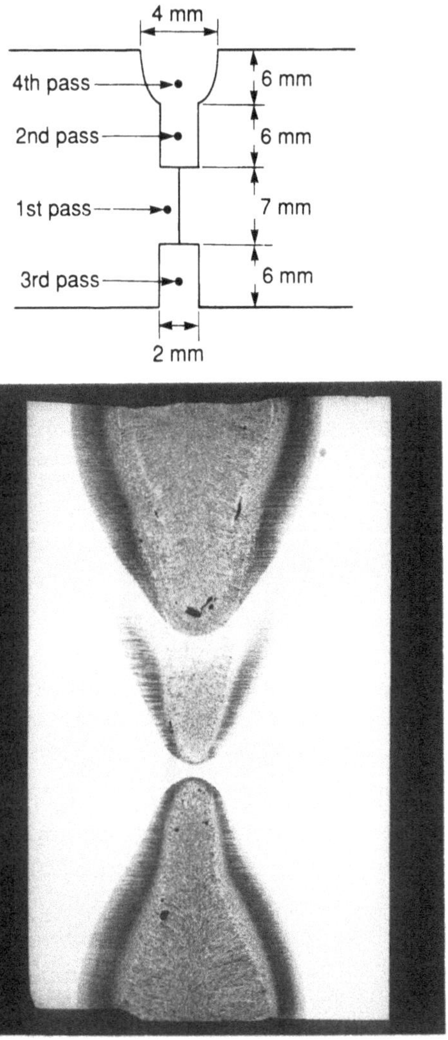


(b)

**8.7** Four pass weld made in 25 mm thick steel using a 5 kW laser:[2] a) Cross section of joint preparation; b) Completed weld. 1st pass: 5 kW, 0.5 m/min, autogenous; 2nd and 3rd passes: 5 kW, 0.35 m/min wire feed speed; 4th pass: 5 kW, 0.30 m/min, 4.75 m/min wire feed speed.

weld pool. Wire pre-heated to red heat enhances the absorption of the laser beam and enables substantially faster welding speeds to be achieved.

## Welding with pre-placed filler materials

The use of pre-placed filler materials has received little attention from researchers, but is worthy of consideration for individual components where weld metal modification is required. Although such consumables may prove more expensive than filler wires, they have a major advantage in that they maintain the simplicity of autogenous welding. Both pre-placed wire and thin sheet materials can be used, provided the welding conditions produce weld zones which completely encompass the filler material and joint walls. Where the pre-placed material is required to prevent weld solidification cracking, such as when welding gear steels, a tightly nipped sheet is preferential to a wire. This is because the wire will have to melt and slump into a recess and will be subjected to more severe contraction stresses than the sheet, which will already be under compression. To weld such items as gear steels, a fully killed steel sheet material with very low carbon, phosphorus and sulphur levels is desirable. For butt welding items such as gears placed on a shaft, the pre-placed sheet material can be conveniently pressed out in a washer form and easily applied during assembly before welding the parts.

## References

1 Sasaki H et al: 'Laser welding with filler wire', Proc 3rd Int Coll 'Welding and melting by electron and laser beam', Lyon, 5–9 September 1983.
2 Watson M N: 'Laser welding of structural steel with wire feed'. Weld Inst Res Report 264/1985.

Chapter 9

# Process performance control and monitoring

Most laser welding applications are on products where a high confidence level in weld quality is necessary. Such products mentioned in this book are fuel filters, saw blades, razors, food mixer whisks and numerous automotive parts. Weld formation and fitness-for-purpose can be carefully established during product development. Moreover, welding procedures and material conditions can and should be laid down and adhered to, to help ensure the required weld quality is achieved in production.

To check the quality of welds in low cost, low production volume components, it is sometimes practical, although wasteful, to withdraw a component after a limited production number and destructively test it, and when an unacceptable failure occurs back-track and test until a sound weld is found. The cause of weld failure is rectified before production is resumed. However, lasers are rarely used to fabricate low cost, low volume production components, because high quality, high volume production welding is usually necessary to achieve realistic laser welding costs and equipment payback times. In these circumstances, regular destructive testing to establish weld quality can prove extremely wasteful in both material and time, and thus very expensive. The alternative is to try and achieve quality assurance by monitoring and maintaining welding conditions within acceptable limits which will produce repeatable welds of acceptable quality and where possible confirm quality by employing non-destructive test (NDT) techniques.

Product liability is important and can have serious legal consequences if failure occurs, especially with respect to the safety of the public. It is, therefore, in the manufacturer's interest to ensure product quality which achieves the required level of safety. Rather than conducting excessive and expensive destructive testing of components regularly taken from the production line, many manufacturers instil confidence and credibility in their products by working to quality assurance standards such as the BS 5750/ISO 9000/EN 29000 series, which is recognised internationally. When working to a quality assurance standard, process control, laser monitoring and NDT techniques can prove beneficial, if not essential. This chapter looks at practical procedures and monitoring techniques which can help achieve quality assurance and control.

The aspects which contribute to producing welds of respectable quality can be grouped under the following headings:

1) Materials;
2) Welding conditions;
3) Periodic post-weld examination;
4) Equipment performance monitoring;
5) In-process weld monitoring.

Items included in 1-3 are very important for reliable weld production and should be adhered to at all times, whereas items included in 4 and 5 are desirable for quality assurance. However, some of these are in their early stage of development, or are experimental and require professional interpretation.

## Materials

Once the materials to be joined have been selected and their weldability established and accepted in relation to the product service requirement, the material manufacturer's specification and chemical analysis should be carefully recorded. Thereafter, all subsequent materials ordered should match the recorded information and the supplier asked to provide a certificate of confirmation. This may sound rather strict, however, small changes in certain material alloying elements can cause serious weld faults, e.g. high carbon, sulphur, phosphorus and lead levels in steel can cause serious weld cracking, likewise unacceptably thick zinc coatings on steel can cause serious weld metal expulsion, blowholes and porosity (see Chapter 4).

The material surface should always be free from oxide scale and other contamination, especially oil and grease. Where present, scale should be removed by turning, milling or grinding and oil and grease removed by chemical cleaning agents, such as Genclene and acetone. Note, cleaning agents which give off vapour must be kept away from the laser beam transmission path to avoid distortion of the laser beam profile; further comments on vapour in the beam path are presented in Chapter 11.

The joint faces should always be close fitting (unless filler wire is being used) and be free from contamination. As laser welding does not use fluxes, cleaning should take place just prior to welding, especially when welding materials which are prone to rapid oxidation, such as titanium; indeed titanium is best 'dry machined' immediately before welding to remove contamination built up during any pre-weld shelf life, moreover, finger grease and oil must be kept well away from the joint faces of titanium.

## Welding conditions

Establishing acceptable welding conditions and tolerance bands (see Chapter 5) is vital and a prerequisite to production welding of any nature. To achieve reliable production welding, the welding equipment to be used and conditions should be clearly recorded on data sheets in relation to the required application and copies issued with any welding work request sheets. In detail, the welding data sheets should identify: the type of laser to be used, workhandling (movement) equipment, focusing optic type (discussed in Chapter 11), the optic focal length, the type of gas shielding device and details of any other necessary anciliary equipment; such as plasma suppression and wire feed devices. The sheet should also clearly define, where possible with practical tolerance bands, the required welding power, speed, type of shielding gas and gas flow rate. The data sheet should also instruct the operator that the stated welding conditions must not be deviated from without official authorisation, and even then any changes must be proven in relation to the weld service requirement, recorded and the welding data sheet updated.

Care should always be exercised in setting the welding conditions with particular attention being given to setting the focus position. All equipment settings, where not continually monitored and controlled, should be checked on a regular basis and ideally, at least before every production shift.

### Laser power

Laser power is normally monitored from a position near the laser cavity; usually by a calorimeter or a thermopile. Some lasers only display the available power while the output beam shutter is closed, the power measuring device doubling as the beam dump, whereas other lasers display the power during the welding operation as well. Monitoring and displaying the beam power during welding is an advantage from a quality assurance standpoint, as the power readout can usually be digitally recorded and stored in relation to the weld number. However, where the laser power is measured close to the laser cavity and the beam is transmitted to the work by a series of mirrors, as can often be the case with $CO_2$ lasers, a small loss in beam power will occur (approximately 3% per mirror) and this loss can gradually increase with the service life of the mirror train. Consequently, the laser power at the work can be different to that monitored at the laser cavity. To avoid this situation it is advisable to measure the laser power just before the focusing optic, using a power probe similar to that in Fig. 9.1. These probes, which are sometimes referred to as 'joule sticks' are simple to use and can measure beam powers up to 10 kW, by placing the absorbency pad in the beam path for as little as 30 seconds, and reading the average power displayed on a calibrated scale. The beam absorbence pad is sized according to the expected beam diameter for a maximum power rating and one should choose a probe with a power

**9.1** A probe for measuring the unfocused laser beam power. These are also known as 'joule sticks' and work by placing the absorbency pad (the lower black disc) in the laser beam path.

rating 15–20% more than the maximum output of the laser on which it is to be used; this is so that readings are not taken close to the ends of the scale. These probes are now commercially available for nearly all continuous wave or pulsed $CO_2$ and Nd:YAG lasers. It is advantageous to purchase a pair of identical probes so that one can be kept to check periodically the performance of the other and also as a back-up in case of failure. The probes are advertised as being hand-held, however, for safety reasons, they are better placed in a retort stand or some other form of mechanical hand and the meter viewed through a safety screen. Laser safety, including safety screens, is discussed in Chapter 12.

### Welding speed

Welding speed is perhaps the easiest parameter to set and check and in many instances can be set with the aid of measured gauge marks and a stop watch. The most accurate method is to hold a simple electronic speed monitoring wheel, Fig. 9.2, against the work and observe the digital readout. These can be built into production equipment and can record during the welding operation. However, like all such equipment they need to be regularly calibrated to ensure accurate readings.

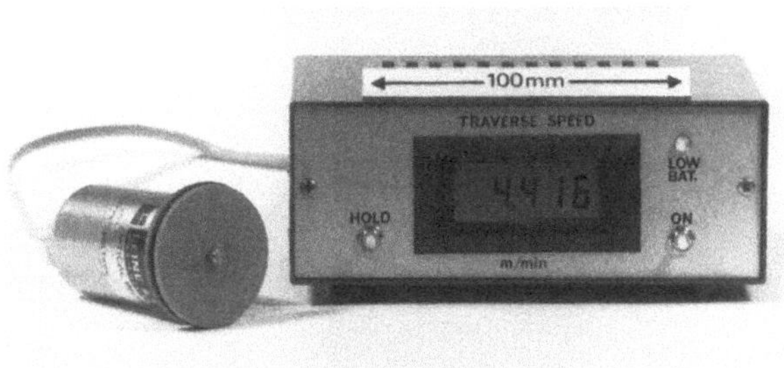


**9.2** An electronic speed monitoring wheel and meter, used for setting and checking the welding speed of the workpiece.

### Gas shielding

Gas shielding can be a sensitive parameter with respect to welding performance and details of equipment and a guide to flow rates is presented in Chapter 5. Ideally, shielding gas should be fed to the dispensing device via a calibrated flow meter fitted with a sensor which will indicate the absence of the correct flow rate. The sensor can also be wired to stop laser operation if the shielding gas supply fails or falls outside the required flow rate limits.

Shielding gas is applied gently to the weld keyhole to avoid disruption of the molten weld metal and therefore care should be taken to ensure that natural drafts or drafts from leaking air or other gas lines, do not impede the flow of shielding gas about the keyhole. This may all sound unlikely, but in fact pneumatic systems are quite common for automatic workhandling and instances have occurred where weld spatter has punctured the plastic pneumatic pipework.

### Focus position

Chapter 5 described in some detail how to find the focused laser beam's minimum waist working position with the aid of test coupons. The same type of test coupon and examination procedure can be used to check the waist (focus) position has not moved. However, as the position is being checked and not established, one must ensure that the surface of the test coupon is in the same position as that of the production workpiece. One may query why the minimum waist position should move once it is set. There are several reasons, the most common is perhaps contamination on the focusing optic, which is usually caused by dust or weld metal vapour. The contamination absorbs beam energy and heats the optic, which in turn expands, distorts and alters the optics focal length. Alternatively, in the case of $CO_2$ lasers, if a mirror in the laser beam transmission train, from the laser to focusing optic, distorts through contamination or lack of water cooling (poor water cooling can also alter the focal length of reflective focusing optics, Chapter 11), the beam propagation angle incident on the optic can change and again alter the optic focal length. Another possibility is if the beam from the laser is caused to move off-axis with the required optical axis of the focusing optic, due to a mirror movement or distortion in the overall system. This possibility can be checked by taking acrylic beam prints, which are described in Chapter 11. Clearly all the above mentioned effects should be checked if the laser beam minimum waist position is found to have moved. High humidity can also affect beam transmission and details are presented in Chapter 11.

If the minimum waist position is found to have moved immediately after changing the focusing optic, for one of the same focal length, then the difference in the minimum waist position is most likely due to the optics manufacturing tolerance. The focal lengths of some manufacturers optics are only to within ± 5% and therefore, for example, two similar 200 mm focal length optics could have, in the worst case, minimum waist positions which are 20 mm apart. Consequently, unless the optics have been accurately manufactured (and they can be if specially requested) the minimum waist position in relation to welding performance should be established as described in Chapter 5 each time a focusing optic is changed.

The aspect of focus position which usually presents the biggest problem and requires special care and the most regular checking, is the focused laser beam alignment with the joint line; unfortunately the narrow laser weld self-imposes a small alignment tolerance (see Chapter 6 for details). Nevertheless, modern laser systems can generally point the laser beam very accurately and with a helium-neon (HeNe) pointing laser (⟨0.5 mW for safety reasons) folded co-axially on the centre line of the path of the welding laser beam, the focused laser beam position can be clearly seen on the work by the tiny (less than 1 mm diameter) red marker dot of the HeNe laser and thus the beam position on the joint line can be gauged before the actual weld is made. When relying on this approach it is necessary of course periodically to check the co-axial alignment of the HeNe laser. This is best done by pulsing the welding laser, at lower power, to produce a fine melt spot on the surface of a test coupon and then ensuring the red HeNe marker dot is concentric with the melt spot by adjusting its position if necessary. The workpiece joint line can then be adjusted in relation to the HeNe marker dot to ensure the welding beam will follow the path of the joint line.

Visually checking the HeNe marker dot in relation to the joint line is usually essential for one-off welds, but impractical for mass produced welds. In the latter case, alignment can be achieved by careful part sizing (tolerancing) the length of one of the two component parts to be joined and placing the carefully sized part firmly against a pre-set backstop in the work holding equipment, thus ensuring the joint always comes in the same place. If this is not possible a joint tracking device is required. Unfortunately such devices are not readily available for laser welding, however, techniques that are promising and could be developed are vision and optical tracking systems.[1]

## Periodic post-weld examination

This chapter so far has looked at the necessary fundamental requirements which, if fulfilled, will provide quality assurance up to the point of making a weld. On the assumption that all the requirements are correctly established in the first instance, and they do not alter during the welding operation, a high level of confidence can be placed in the correct formation of the resultant weld. However, one cannot be 100% sure that things will not change during the welding operation. For example, it is possible that during a long welding cycle, a slightly contaminated focusing optic, or a poorly water cooled mirror could thermally distort and reduce the focused power density and cause an unacceptable loss of weld penetration. Regular weld formation checks are therefore necessary.

Assuming the welding application is proven, sometimes a simple visual examination can be sufficient to confirm an acceptable weld, for example, a lap seam weld which has clearly penetrated both sheets, with no obvious sheet separation and exhibits clean smooth top and underbead weld profiles, can generally safely be accepted. A lap weld, however, that lacks these characteristics and has obvious weld blowholes is unacceptable and clearly indicates something has gone wrong during the welding operation. Where the joint configuration or construction does not permit adequate visual inspection of the weld, it is sometimes possible to use an ultrasonic or X-ray technique to check the weld position and depth (the X-ray technique is expensive and time consuming and therefore not always suitable as a production line check). The ultrasonic pulse echo technique[2] can be good and its method of finding a weld that has not penetrated down or which has missed the joint line, is shown schematically in Fig. 9.3. Unfortunately, apart from butt joints in plate and tube, many laser welded components do not lend themselves to ultrasonic testing; usually because they deny access and ease of coupling of the ultrasonic probes.

Sometimes it is possible to check weld acceptability by proof testing, for example performing a hydraulic pressure test on a welded cylinder, or a domestic heating radiator, where the item is stressed to an expected safe value, which is beyond the item's service requirement. Thus if the weld does not fail, it can be assumed that it will be safe for its normal working requirement. There are numerous proof tests and they range from simple mechanical loading to complex high vacuum leak tests; the latter are quite common for checking the welds used to attach and hermetically seal the lids of micro-circuit packages. Proof

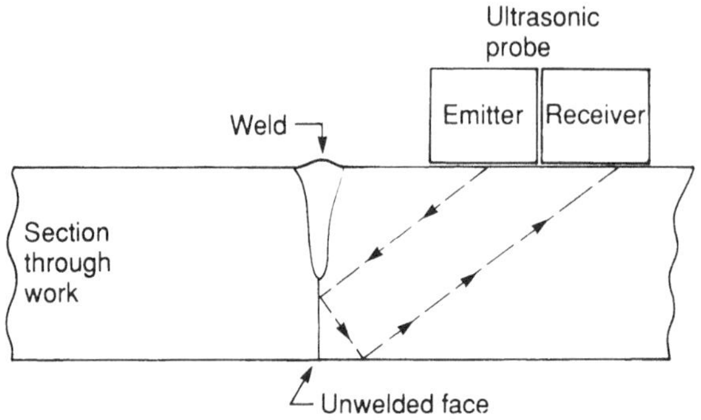


**9.3** The principle of a pulse echo, non-destructive, weld monitor. A short burst of high frequency sound, usually in the range of 1 MHz to 5 MHz, travels through the work and on striking an interface or discontinuity is reflected. By careful positioning, the probe can transmit to and receive back a sound wave reflected from a selected target as shown above. In reality, had the weld in the illustration penetrated below the position where the emitted wave meets the interface, the wave would have travelled on through the solid material and not been reflected back to the receiver, thus indicating the weld depth had passed the probe monitoring point.

testing, however, should be treated with caution and selected with care as the chosen test may not cover all the service requirements. For example, a good hermetic weld, found by vacuum leak testing can have poor mechanical strength or vice versa.

Welds on a component that cannot be checked by the non-destructive and proof test methods described so far, usually have to be tested in a manner which destroys the component. The type and choice of test depends on the component part and its service requirement, typical tests are peel, tensile, shear and macro-section. The most difficult decision to make is what frequency of testing is necessary. The decision initially has to be taken in relation to the considered overall reliability of the welding operation; particularly equipment performance, component value and production numbers. Equipment performance is, of course, closely related to design and maintenance (Chapter 13). Moreover, the satisfactory operating period of certain items, before servicing is necessary, often has to be established by experience. For example, the frequency of cleaning the focusing optics can depend on the application and on whether the metal vapour soot and weld spatter from tooling is removed before it flakes off and falls into the weld zone.

Component value is perhaps the biggest drawback to taking samples for destructive testing, because the welding operation can often be the final stage in the production of some components and by the time the welding operation takes place the total manufacturing cost is already very high, as when sealing a heart pacemaker case. In such situations, often the best answer is to introduce and test a dummy component which exactly represents the material and joint configuration. Having reduced the component value problem, one is still left with the most difficult question, the frequency of the destructive test. There is no simple answer to this question and, in general, destructive test schedules are developed from operating experience gained by taking cautious steps and careful documentation. That is, building case histories, based on gradually extending the production number in relation to equipment service schedules, before destructive test sampling takes place. In this manner, one may start by testing one weld in ten and if acceptable make 50 before the next test and ultimately establishing that, by careful control of all the welding aspects, it is only necessary to test destructively one weld in 100 or perhaps more.

## Equipment performance monitoring

When an inadequately formed weld is found during periodic testing and all the requirements of the welding data sheet appear to have been met, the cause can usually be traced to the laser operation or the beam transmission system. One may question what could be wrong with the laser if the power readout is correct. The first possibility is of course a fault in the power meter. Should this not be the case and the focusing optics are known to be sound, the cause of poor welding performance is likely to be a deterioration in beam quality, either in the laser cavity or during transmission to the focusing optic. It is possible, especially with $CO_2$ lasers, that an optical or lasing medium fault in the laser cavity can produce the required laser power, but with a distorted or different beam mode structure (Chapter 1), thus affecting the focus spot size and consequently the welding power density. Ageing of Nd:YAG flash lamps can also create a similar problem.

To examine the cross sectional shape of laser beams and their mode structures, several types of laser beam analyser have been developed. These devices can be used to check beam characteristics during the welding operation and thus provide a method of quality assurance with respect to the laser beam. Moreover, if the beam characteristics fall outside an acceptable requirement, the signal can be used to shut the laser down before any further welds are made.

If a laser beam analyser is used on a production welding machine, it is suggested that it should be positioned in two particular places within the beam transmission system. First, to monitor the laser beam during the welding operation, it is suggested the analyser be placed before and near the focusing optic, thus monitoring the beam at the end of its unfocused transmission. Then, second, if a deterioration in the displayed beam profile occurs, the analyser can easily be moved to a position at the beam output point of the laser. The two positions will then enable one to determine if the laser beam profile deterioration has occurred during beam transmission, between the laser and focusing optic, or within the laser cavity. One should note that the diameter and profile will normally be different at each of the two suggested examination positions, and therefore one should have picture records, taken when the system is first established, to indicate what beam diameter and profile is expected at each position under normal circumstances.

Laser beam analysers generally take the form of a rectangular box, Fig. 9.4, with a hole from one side to the other, through which the laser beam can pass. Within the hole most analysers have a needle-like beam sampling rotor, which passes at high speed through the laser beam and at right angles to the beam axis. The variation in light intensity sampled by the rotor, as it sweeps through the cross section of the beam, is transmitted to a sensor system and converted to an electronic signal, which enables a picture of the beam power density profile to be built up.

Some analysers display only a two dimensional picture of the beam profile[3] and of this type some take two sections through the beam at approximately 90° to one another. The two dimensional pictures are displayed on an oscilloscope screen. More recent analysers[4] have the ability to display, with the aid of computer graphics, three dimensional isometric pictures, as shown in Fig. 9.5. These equipments are expensive and require expert operation and sometimes picture interpretation.

Laser beam analysers can also usually be positioned (prior to a welding operation) to look at the size of the focus spot. However, from the production welding engineer's standpoint this facility is perhaps academic, because the focus spot size formed during the welding operation and amid metal vapour and a plasma may be quite different from a laser beam focused in air. Furthermore, there does not appear to be a calibrated test which would show the focused spot dimensions displayed by present analysers are in fact accurate.

**9.4** A laser beam analyser. The simple box shape enables the analyser to be easily installed, in drawer and chest form in a laser beam transmission system.

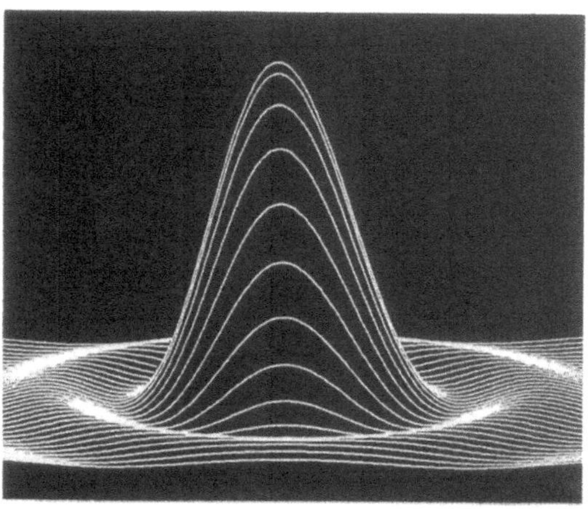

**9.5** An example of a three dimensional computer drawn picture of the power intensity distribution through a transverse cross section of a laser beam. The picture has been built on signals recorded from a laser beam analyser.

## In-process weld monitoring

So far weld quality assurance by attention to materials, welding conditions and pre-focus laser beam characteristic monitoring has been described. Despite all these good practical steps, quality assurance cannot be guaranteed for several reasons. For example, a fault occurring in the focusing optic, such as thermal distortion or a cracked lens, will distort the focus spot and change the focused laser power density. Alternatively, imperfections in the material or unforeseen contamination such as general factory or welding dirt or moisture entering the weld zone, can affect the absorption of the laser beam energy by the work and hence produce an unacceptable weld formation.

To achieve as near 100% quality assurance as possible, a sensor is required which would confirm an acceptable weld formation, encompassing the required joint area, has been achieved. Such a device for production line use is of course every welding and production engineer's dream and for which researchers continue to strive. For laser welding two approaches have met with some success, but are limited in application. They are ultrasonic monitoring, as described earlier for post-weld inspection and ultraviolet (UV) monitoring. Thermal methods have been successfully used for laser soft soldering, but are unsuitable for welding.

Ultrasonic monitoring, as stated earlier, indicates weld depth in relation to joint line coverage. It operates typically 30 mm behind the keyhole to protect the ultrasonic probes from thermal and weld spatter damage. The latter is unfortunately a drawback to developing the technique for a closed loop feed-back control of weld depth by adjusting welding power and/or speed. Nevertheless, if small ultrasonic probes could be developed to access joints and also achieve the required surface contact, they could play a much wider and major quality assurance role; especially in the manufacture of partial penetration and blind welds.

Ultraviolet monitoring, Fig. 9.6, indicates the transfer of laser beam energy into the work by monitoring the average UV emission from the plasma cloud above the weld keyhole. It does not evaluate weld depth or indicate the weld has encompassed the joint line. However, when correctly installed and set up it can detect changes in the focusing system, gas shielding, welding power and welding speed by the changes these aspects produce in the plasma formation and resultant UV emission. Consequently, despite obvious parameter changes, such as power and speed, which can be monitored elsewhere, it can detect a contaminated or cracked focusing optic or a change in the component material sur-

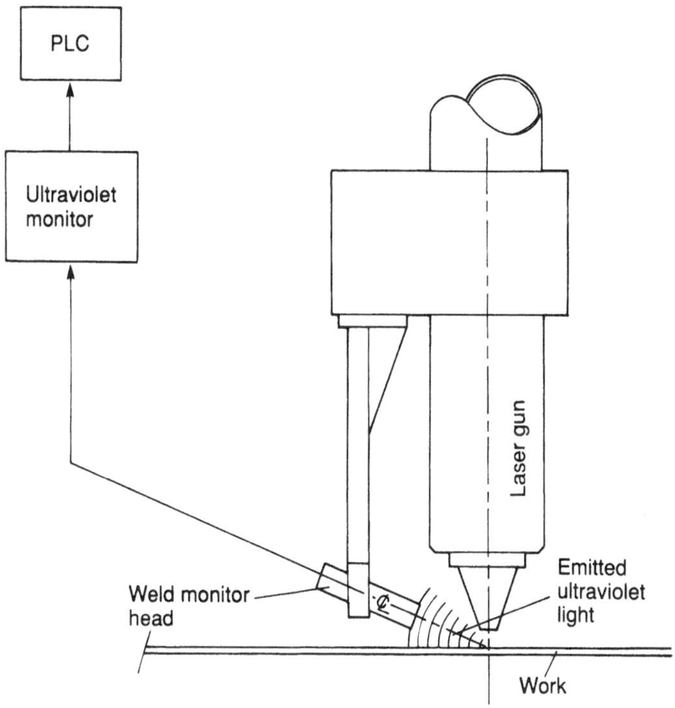


9.6 General arrangement of an ultraviolet weld monitoring head in relation to the laser gun and workpiece.

face condition, due to fit-up problems or surface contamination. Being able to monitor the focusing aspect is perhaps its most important feature, as spotting a focusing fault is paramount to the laser welding technique. Moreover, to provide a signal to shut down an automatic welding machine will eliminate many reject parts and the anxiety of not knowing where in the production run the fault first occurred.

To use UV monitoring successfully one first has to optimise the welding conditions and their tolerances and then measure the UV output from the plasma produced. From these measurements the acceptable UV working range is established.

The plasma and hence UV signal is not stable, in fact there is a strong signal modulation due to the formation and collapse of plasma balls in quick succession. The UV signal is therefore examined over a short time-period and based on an average signal. The signal is usually digitised to enable a comparison with the present go–no go values. Setting up a UV monitoring system requires careful experimentation to ensure the acceptance limits are correct and thereafter regular periodic calibration and testing by imposing deliberate faults.

## References

1 Fryatt A: 'Sensor technology launches robotic arc welding'. *The Fabricator*, May/June 1988.
2 Kroutkramer J and Kroutkramer H: 'Ultrasonic testing of materials'. Springer-Verlag, ISBN 0-387-51231-4.
3 Lim G C and Steen W M: 'Measurement of the temporal and spatial power distribution of a high power $CO_2$ laser beam'. *Optics and Laser Technology*, June 1982, pp 149–153.
4 Laserscope UFF 100, Prometec GmbH, Julicher Str 338, D-5100 Aachen.

Chapter 10

# Work handling, laser beam manipulation and process automation

Successful laser welding depends on the way the work is held and manipulated beneath the laser beam, or the beam manipulated about the work. The work has to be held securely to ensure that the required joint line relationship with the laser beam is achieved. The joint also often requires a compressive axial holding force across itself to avoid opening as the preceding weld metal, of a continuous weld, solidifies under the associated high contraction forces. The workpiece or laser gun has to be manipulated accurately to ensure the laser beam focus position is maintained and that the beam correctly encompasses the joint. Furthermore, for continuous welds (welds other than spot welds) control of the welding speed is important to maintain the weld energy input per unit length of weld and hence achieve a consistent weld width and depth profile.

To give guidance for achieving the aforementioned, this chapter describes the various work holding and work or beam manipulation methods (shown schematically in Fig. 10.1) which will cover most laser welding applications. Methods of welding automation which increase production and maximise the effective use of laser power are outlined and finally, important considerations when making butt joints in plate materials are presented.

## Workpiece manipulators

Rotary workpiece manipulators are the most simple and common, as the majority of mass produced laser welds are made in some form of circular butt or lap joint (see Chapter 2, Fig. 2.8 for a typical example). Rotary manipulators are available from many welding equipment suppliers, although for laser welding the speed range may have to be increased. The latter requirement also sometimes necessitates a larger drive motor to produce the torque necessary for the component weight.

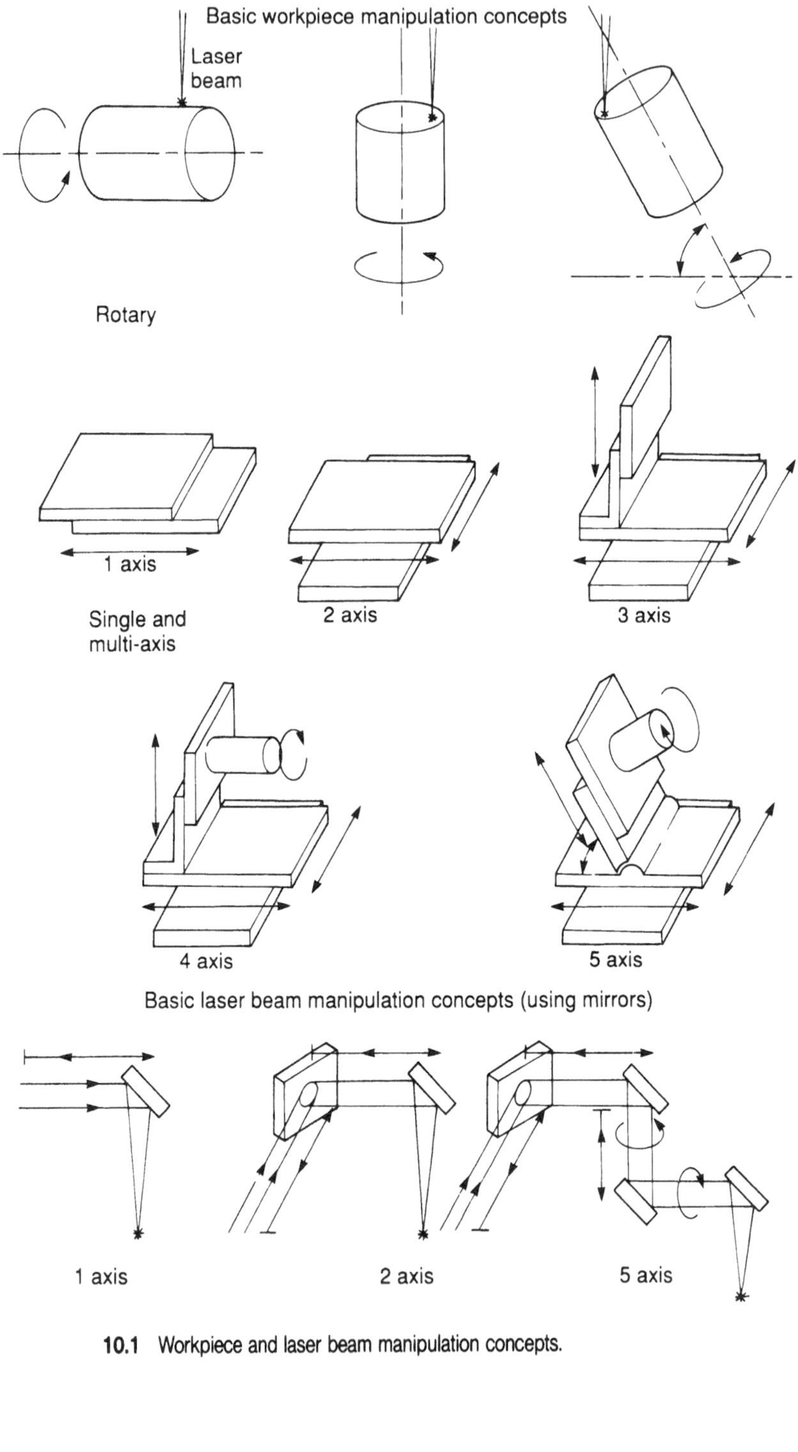


**10.1** Workpiece and laser beam manipulation concepts.

For small production and one-off work, a small lathe can be used to rotate the work and where necessary an adaption fitted to the tailstock can apply an axial compressive force on the joint faces. On the other hand, a purpose-built rotary table can usually operate with its rotating axis positioned anywhere between the vertical and horizontal and thus offer more flexibility to component variations and laser beam access.

Workpiece location on rotary manipulators can be achieved by face plate mounted machine tool collet holders, chucks, general engineering clamps and draw-bars. Internal and external work location spigots, with some form of work locking device, can also be used to provide accurate and quick positioning. Care must be taken when mounting workpieces requiring butt or lapp joints on rotary manipulators to ensure the joint line rotates concentrically with the axis of rotation, otherwise the work surface will move in and out of the laser beam focus position as the work rotates. Also, to ensure the laser beam remains on the joint line, the line must be positioned square to the axis of rotation.

Much larger and heavier components can be handled by rotary tables than by linear movement tables, because control of acceleration, deceleration and the combined inertia of the rotating table and workpiece are not necessarily dictated by a maximum distance of movement and cycle time; the rotary table can rotate as many times as is necessary to achieve the required action. Consequently, quite large components can be handled on rotary tables before it becomes necessary to consider manipulating the laser beam, with its added complexity, about the workpiece. For an example of rotating a large component, see Chapter 3, Fig. 3.19 which shows a 2.5 m diameter stainless steel hoop being rotated in front of a laser gun.

To make straight welds on flat surfaces up to 1 m long, a wide choice of commercial work tables is available from numerous suppliers. Most of the tables capable of 1 m movement operate at welding speeds up to and sometimes in excess of 7 m/min. Generally driven by DC servo or linear motors with encoder speed control, they also have relatively high rates of acceleration and deceleration. Tables requiring movement greater than 1 m can also have these attributes, but usually have to be made to special order. Many of the commercially available tables are designed for a wide range of industrial applications and consequently not all are suited to welding operations. Therefore, when choosing a table for welding, care must be taken to ensure that it will comfortably handle the workpiece size and weight; it is sufficiently robust for the production environment and the slideways are well protected, by non-combustible guards, from weld spatter.

To make non-linear welds, computer controlled, X-Y tables are available, see Fig. 10.2. These comprise two linear tables, one mounted on top of the other, with their axis of movement at right angles. As with the single axis linear tables, X-Y tables with movements greater than 1 m are usually only made to order.

Whether the work tables are linear or X-Y, they must be mounted on a solid pedestal base, which can react to the combined inertia of the moving table and work as they come to rest or change direction. Tables with movements up to 1 m are rarely supplied with a pedestal and in such cases, old milling machine beds, as shown in Fig. 10.2 can make ideal pedestal bases. Also, if the table is mounted on the milling machine bed knee, it will provide an accurate Z movement (vertical up and down) control, which will enable the focused laser beam position to be simply adjusted and accurately set in relation to the workpiece surface.

Many of the commercially available work tables are not fitted with adequate T slots and in a job shop environment it is often necessary to have a T slotted table top specially made to allow different shape workpieces to be handled. On a cautionary note, when fitting a T slotted top, one must remember they are heavy and that the manufacturer's stated safe table working load for the X-Y table will in fact be the stated weight less the weight of the T slotted top.

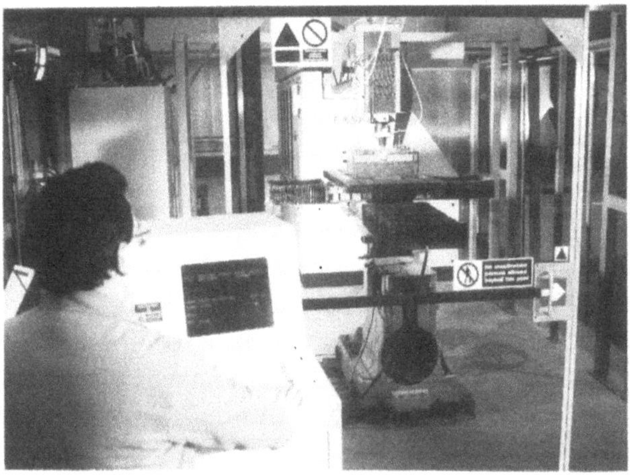

**10.2** A computer controlled X-Y table mounted on a pedestal made from an old milling machine bed.

The use of linear and X-Y tables for moving large components, which invariably also require long weld lengths, can pose numerous problems. For example, the acceleration distance the table has to travel before it reaches the required welding speed, and likewise the distance for deceleration, have to be added to the travel distance required for the weld length. Consequently, if high welding speeds are required, a table with a total travel distance somewhat longer than the weld length will be necessary. Moreover, large tables and large components exacerbate the problem of reacting to the inertia developed during table movement. Also, when using large linear and X-Y tables, an often unforeseen disadvantage is that they require a large floor area to allow for the table travel distance. For example, a $2 \times 2$ m, X-Y table with 2 m travel in each axis, will require a floor area of at least 16 $m^2$, plus sufficient area for access.

## Laser beam manipulation

The answer to the problems associated with large tables and workpieces is to move the laser beam about the work, or if possible, compromise by moving both the beam and the work. Linear movements of the laser gun, up to 2 or 3 m and without a laser beam collimator can be achieved quite easily by employing an inverted precision ground track and a saddle carriage for the laser gun, supported beneath a gantry, as illustrated in Fig. 10.3. The carriage can be moved by a motorised lead screw, or even by a variable high speed motor and high reduction gear box winching the saddle carriage via a wire cable. In the latter case, to avoid snatching and an erratic action, the cable can be operated in a push-pull regime, by passing it round helical grooved winding drums placed at each end of the gantry.

### $CO_2$ lasers

When used with a $CO_2$ laser, the simple linear movement laser gun gantry system does not necessarily introduce any more beam fold mirrors into the beam transmission path than are used by a work station with a fixed laser gun (the advantage of this is described shortly). Furthermore, without introducing any more mirrors, a Z movement axis for the gun can be applied. Combining these two axes of gun movement with a single linear table and driving each by computer control will

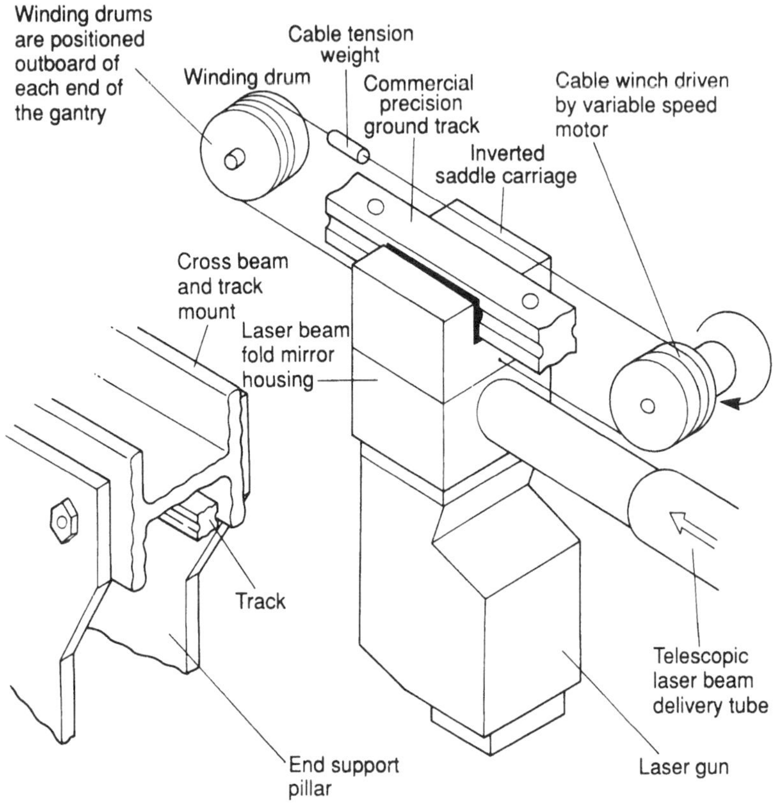


**10.3** General arrangement for a single axis linear gantry, which can be built in-house using commercially available components. This is based on an I section cross beam supported by an A frame at each end.

provide a basic programmable three axis robotic system. If two additional reflective optics are used, one of which could focus the $CO_2$ laser beam, then 5 axis beam/workpiece manipulation can be achieved, as shown schematically in Fig. 10.4. Should the work be too large for a linear table, an X-Y gantry robot, Fig. 10.5 can be used in conjunction with the other gun movements. Large gantries of this type, however, have to be precision made and often temperature-controlled to maintain the necessary very high accuracy of optical alignment through the beam transmission mirror train. Also, as the laser gun moves about its

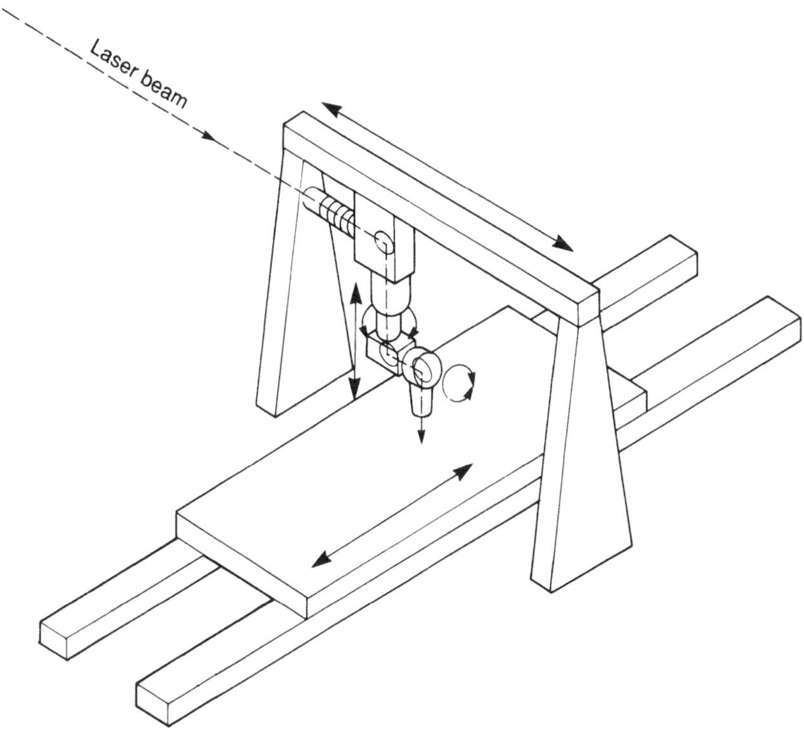


10.4 General arrangement of a gantry with 4 axis laser gun movement, combined with a single axis linear table and thus providing a 5 axis robotic system.

working envelope, the laser beam path length continually alters. Where the movements amount to several metres, a laser beam collimator is necessary to maintain the laser beam diameter within acceptable limits. Otherwise, the focus f number will continually change and simultaneously alter the laser beam focus spot size, thus causing unwanted changes in the welding power density.

Multi-axis gantries operated in conjunction with $CO_2$ lasers, apart from being very expensive, lose a substantial amount of laser power during beam transmission and, therefore, a higher power laser is required than would be necessary for a simple welding station where the work is moved. Each mirror in a $CO_2$ laser beam transmission train absorbs approximately 3 to 5% of the laser power, depending on the mirror condition. A gantry with a 5 axis laser gun movement usually has at least 6 mirrors within the beam transmission train. In such a system employing a laser working power of 6 kW, the laser power at

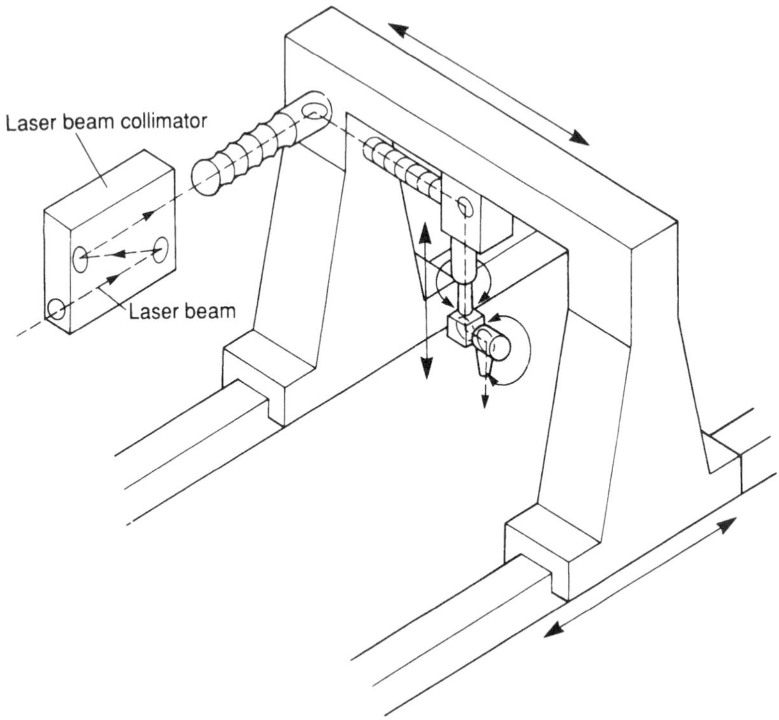


**10.5** General arrangement of a 5 axis gantry based robot for manipulating the laser gun about large stationary workpieces.

the work would only be in the order of 4.4 to 5 kW; a potential power loss of 17 to 27%.

Before leaving the subject of $CO_2$ laser beam manipulation and work handling equipment it is worth noting that some ingenious 'laser cells' have been developed, which make versatile workstations, especially for variable job shop work. A commercially available system, which combines 5 axis beam movement with 3 axis workpiece movement, is shown in Fig. 10.6.

## Nd:YAG lasers

Nd:YAG lasers have a distinct advantage over $CO_2$ lasers when it comes to beam manipulation about the work, and also when one laser is required to work multiple workstations. The advantage is that the

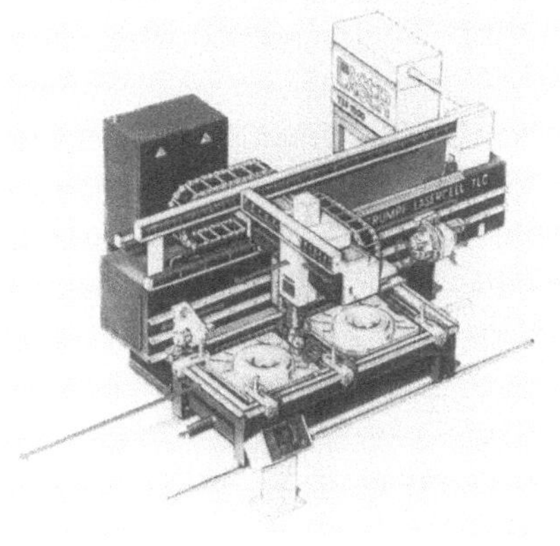

10.6 A versatile laser work station, which can handle many different component shapes and joint configurations by calling on combinations of work and/or laser beam movements.

1.06 μm wave length Ng:YAG laser beam can be transmitted down a fibre optic with no appreciable loss of power; the power loss aspects of fibre optics are described in more detail in Chapter 11. This ability means the light can travel directly from the laser via a flexible cable to a laser gun mounted on an articulated wrist of a robot arm, Fig. 10.7 or a gantry without a significant loss of laser power. At present, the commercially available fibre optics only operate at laser powers of little more than 1 kW. Fibres are being developed to handle the beams from the multi-kilowatt Nd:YAG lasers currently under development. If these developments are successful, then laser beam manipulation about large workpieces requiring deep single pass welds (greater than 10 mm) will be considerably simplified and allow much wider areas of application than are offered by the optically delicate and inefficient power transmission $CO_2$ laser gantry systems.

## Process automation

Both Nd:YAG and $CO_2$ lasers are ideally suited to production automation. Laser guns can be inserted in production lines, provided safety screening is present to protect the line personnel; safety screening is discussed in Chapter 12. Moreover, the laser can be positioned some distance from the production line and the laser beam piped to it. One laser can operate multiple workstations by switching the laser beam from station to station, thus, whilst welding in one station, part loading

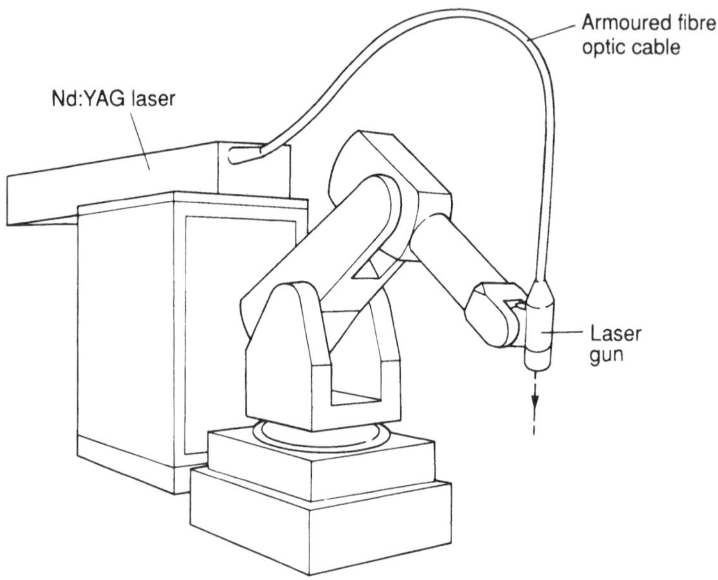


**10.7** The optical simplicity offered by an optical fibre for transmitting an Nd:YAG laser beam to a multi-axis moving laser gun; compare with Fig. 10.6.

and unloading can take place in other stations. On the other hand, several very different stations can time share one laser. Beam switching maximises the laser use and with many types of $CO_2$ laser, especially those employing aero-windows, much more efficient use of the laser beam is made, as the time the beam spends in the calorimeter/beam dump, between welding operations, is considerably reduced. The latter point is a very important factor in relation to the laser operating cost per weld.

The multi-mode beam from an Nd:YAG laser can be split, Fig. 10.8(a), by inserting staggered beam folding mirrors into and across the beam path. Thus the beam splitting system, in conjunction with a fibre optic beam delivery system, can make several welds simultaneously in one or more workstations. Alternatively, the beam can be sequentially switched, Fig. 10.8(b) to different points, often up to 30 m away. In fact there are industrial micro-spot welding systems where the beam is switched among eight workstations at up to 40 times per second.

Where the Nd:YAG laser beam is split, the odd shape of each individual cross section is homogenised into a focusable form by transmitting it through the fibre optic. This action is not available to $CO_2$ laser

(a)

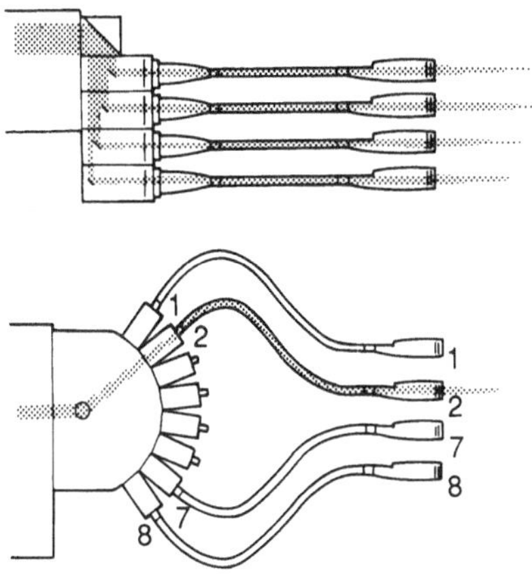


(b)

**10.8** Laser beam sharing system which can be used with Nd:YAG lasers: a) Splitting a single beam; b) Beam switching between stations. Both systems use optical fibres for beam transmission.

beams, because at present there are no fibres which are suitable for transmitting the 10.6 μm wave length. Nevertheless, $CO_2$ laser beams which do not exceed a few kilowatts in power can be split by placing partially transmitting beam folding mirrors/windows in the beam path, see Fig. 10.9. The amount of beam split off at each mirror is controlled by the percentage of reflective coating uniformally dispersed over the optic face. Unfortunately, this form of beam splitting cannot handle high laser powers, because they cause thermal lensing of the optical material and hence variable beam distortion in relation to transmission power level. It should also be noted that at practical working powers, the beam splitting optics are prone to premature thermal distortion if slightly contaminated by airborne particles and therefore a protection system and regular cleaning are necessary. These two aspects are covered in the next chapter.

Sequentially switching a $CO_2$ laser to different work stations can be easily achieved, at all laser powers, by inserting fully reflecting fold mirrors into the beam path, Fig. 10.10. Such systems are well proven and quite common on production lines for welding automotive parts.

## Workpiece clamping and weld underbead spatter channels

Although some general comments have already been made during the course of this book, the importance of good work clamping needs to be

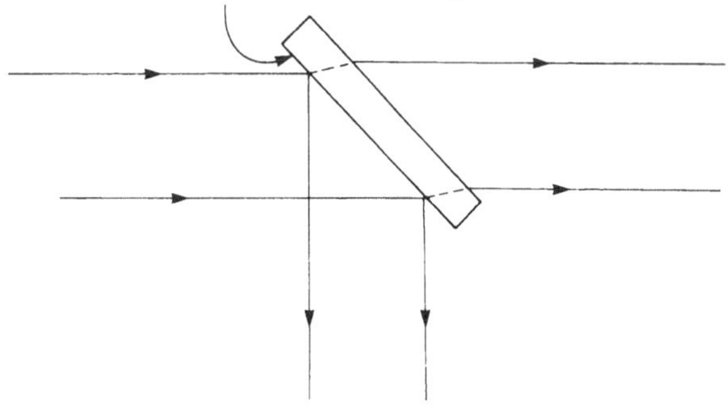


10.9 Beam splitting with a $CO_2$ laser beam by using a transmissive optical flat, with a carefully dispersed reflective coating on one side. Note, due to the refractive index of the optic material, its thickness and angle of incidence to the beam, the beam transmitting through the optic will be marginally displaced from its original axis.

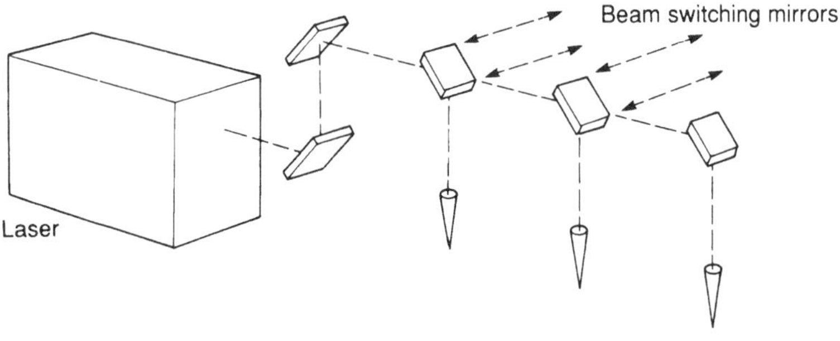


10.10 $CO_2$ laser beam time sharing by inserting fully reflecting mirrors sequentially into the laser beam path.

emphasised. Component distortion and hence joint line movement are not normally very serious problems when welding precision made circular parts, however, when making long butt and lap joints in most other sections, especially flat ones, serious joint opening problems can occur due to contraction forces during weld solidification. Moreover, where low welding speeds are employed the heat transfer into sheet and plate materials can cause buckling due to thermal expansion and local stress relieving. For long butt and lap joints in thin sheet steel, clamping bars, placed close to the weld line are essential (see Chapter 6). These can take the form of multiple screw-down fingers, which apply a uniform pressure, or alternatively a flexible steel bar held down by a series of closely spaced hydraulic or pneumatic cylinders, as shown schematically in Fig. 10.11. Such systems are commonly used to hold down thin sheets when making long butt joints between either laser cut or precision guillotined edges.

When making butt joints in plate materials, especially in thicknesses above 6 mm, some form of axial clamping to compress the joint faces is desirable to stop the joint opening as the preceding weld shrinks, see Fig. 10.12. This effect unfortunately happens very easily and can also be caused when making the initial joint tacking welds, as described in Chapter 7, if the parts are not properly clamped. The clamping equipment for pushing plates together and holding them in position to avoid vertical and lateral movements needs to be robust and well made to be effective. Care taken here can save the often difficult task of trying to reclaim an opened joint by using filler wire, or forcing a suitable material in the opening and making a further weld pass in the hope that the required weld penetration width and depth will be achieved. Unfortunately, there is no one design of clamping system which will

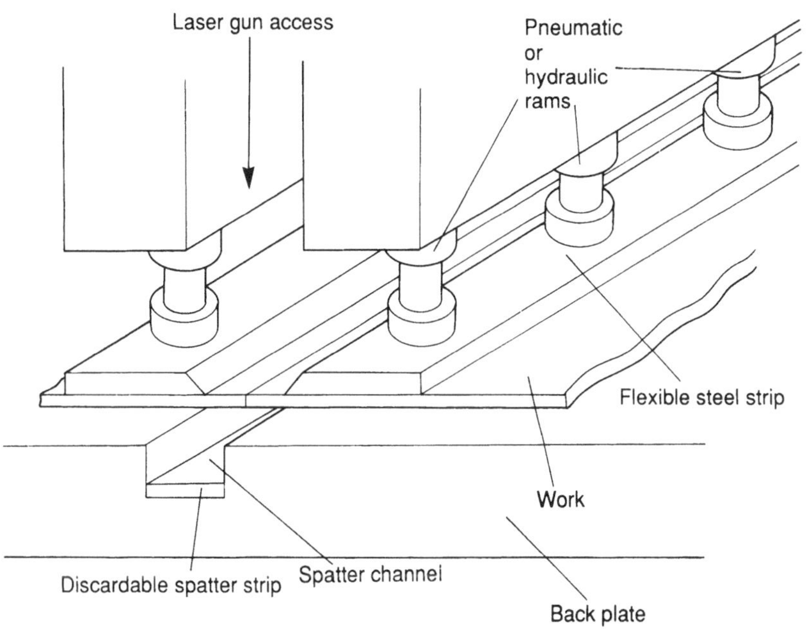


**10.11** A common method for rigidly and accurately clamping long lap and butt joints in thin sheet materials.

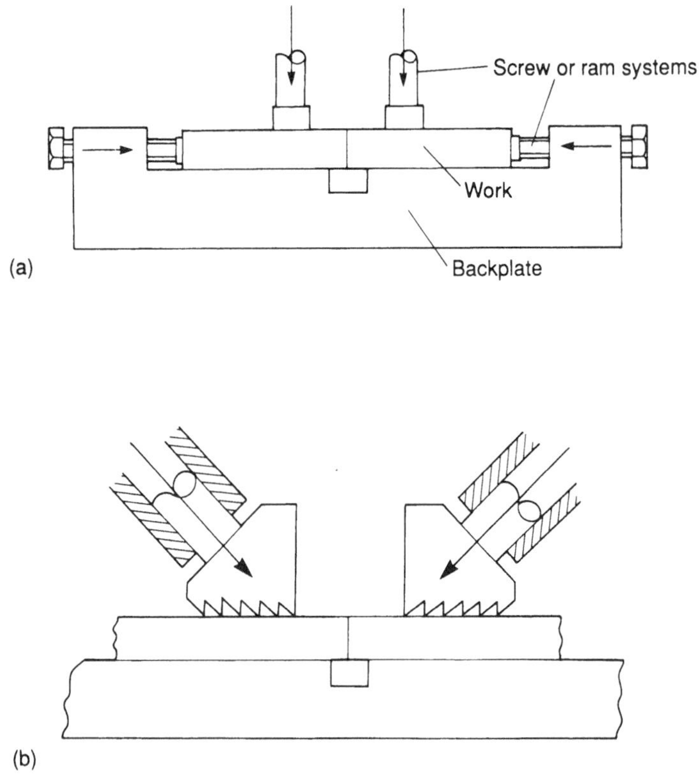


**10.12** Methods of clamping butt joints firmly together showing (a) where it is practical to apply loading on the edges of the workpieces and (b) where loading on the edges is not possible. This approach requires high clamping forces and causes surface marking.

cover all eventualities, each system has to be designed in relation to the component to be welded. Nevertheless, if the above comments are heeded and adequate clamping precautions are taken the problems of joint opening and surface mismatch should never arise.

An annoying and often unforeseen problem when making joints that are clamped against a supporting backplate, is weld damage and contamination by back spatter. This occurs because either the necessary channel in the backplate running beneath the weld line, see Fig. 10.11, is not deep enough, or after a succession of welds it has become filled with spatter and consequently back spatter through the weld keyhole occurs because its normal escape route is suppressed.

A thin throw away strip similar to the material being welded can often be placed in the channel and discarded before an unacceptable level of spatter builds up. However, such strips need to be held in position, otherwise they can bow up and become incorporated in the weld. Some workers advocate a copper strip in the belief that it will reflect laser beam energy back into the keyhole. Experience indicates this has no effect, indeed, it only offers the possibility of damaging the weld metal with copper vapour if melted. Likewise, bare channels in light-weight aluminium alloy back plates should be avoided when welding sensitive materials. Nevertheless, when welding materials sensitive to oxidation an underbead gas shield can often be applied via the channel in the backplate. In such circumstances one end of the channel must be left open, otherwise the underbead shielding gas could pressurise the weld keyhole and cause expulsion of metal or weld porosity.

Chapter 11

# Optical equipment

Optics are the essence of laser machine and laser processing technology and a highly technical subject in themselves. In the early days of laser welding one approached a general optics manufacturer, described what one wanted to do and the manufacturer attempted to supply an optic which would meet the requirement. However, because little knowledge of the materials and their service requirements was available, the whole exercise was often a hit and miss affair. Today there are companies who specialise in optics specifically designed for laser welding and who work closely with users to perfect continually their product[1-5] and when necessary should always be consulted. Before consultation, however, the user needs some fundamental understanding in order to be conversant. To assist the beginner, this chapter introduces the basic different optic types, materials and optical systems used in laser welding.

The types of optics used in laser resonators are not discussed here, they are the domain of the machine manufacturer and specially trained service personnel. Passing reference is, however, made to laser output windows with respect to potential laser beam transmission problems.

## Focusing optics

A guide to the practical aspects of focusing optics, e.g. spot size, depth of focus, focusing f number, focus position, etc, is presented in Chapter 5. Here the different types of focusing optics and optical materials are described.

Focusing optics fall into two distinct groups: transmissive optics (lenses) and reflective optics (shaped mirrors).

## Transmissive focusing optics

In Chapter 1 we saw how most of the sun's light wave lengths will pass through conventional soda glass. Carefully refined glass (borosilicate crown glass, BK-7, being a typical example) is used to make the lenses for focusing the 1.06 μm light wave length from Nd:YAG lasers. Unfortunately, glass is opaque to the ten times longer wave length of $CO_2$ laser light and therefore special and much more expensive materials have to be used. Although numerous materials have been experimented with for use with $CO_2$ laser light, only two have so far proved practical for transmitting and focusing the powers necessary for welding. These are zinc-selenide (ZnSe) and potassium-chloride (KCl), which are able to transmit $CO_2$ laser light efficiently due to their very low absorption and high transmittance characteristics.

ZnSe has a refractive index of 2.4 at the wave length of 10.6 μm and KCl about 1.5. The refractive index is important because it determines the amount the laser beam light waves are bent on passing in and out of the optical material. The material refractive index is therefore carefully considered by optic manufacturers when establishing the lens curvature for a given focal length.

KCl lenses are five to ten times cheaper than ZnSe lenses and therefore for applications such as research where the lens is likely to be abused and quickly damaged, they have an advantage. On the other hand, KCl has two disadvantages when compared with ZnSe. Firstly, KCl is hygroscopic (it absorbs moisture) and being a salt material it will dissolve and cloud if used in an area of high humidity and thus lose its focusing and transmission capability. Secondly, it is difficult to coat with an anti-reflection coating and therefore does not have the light transmittance and focusing ability of ZnSe.

Experience has shown both ZnSe and KCl, providing they are kept clean, will satisfactorily handle laser powers up to 5 kW, but much above this level they are prone to thermal distortion and fracture; usually through only slight contamination causing absorption leading to thermal runaway and subsequent failure.

Two main types of lens shape are commonly used for laser welding, they are the plano-convex and the positive meniscus, which are shown in cross section in Fig. 11.1. Plano-convex lenses are the less expensive and are suitable for most applications where focus spot size and spherical aberration is not critical (spherical aberration is described shortly).

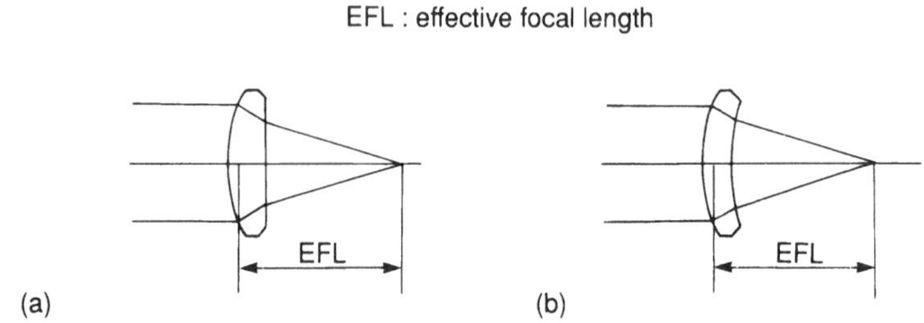


11.1 Cross sections of plano-convex and positive meniscus lenses: a) Plano-convex lenses have a curve on one side only - the steeper the curve, the shorter the focal length; b) Positive meniscus lenses have two curved surfaces - the steeper the curves, the shorter the focal length.

Positive meniscus lenses are designed to minimise spherical aberration and thus they provide a smaller spot size for a given focal length and therefore a higher power density at the workpiece.

The positive meniscus lens is recommended for welding operations where the f number is likely to be f3 or less. For larger f numbers the plano-convex lens is generally quite satisfactory. Plano-convex lenses do not perform well at f3 and below because their short focal length dictates that they have a steeper curve (tighter radius) on their back face which gives a subtantial change in lens cross section over the diameter of the lens. This results in unequal thermal expansion across the lens, plus substantial increased spherical aberration, which together alter the focal length and scramble the focus spot size by altering the paths of the converging rays. The more uniform cross section of the meniscus lens and the more shallow back face curve can almost eliminate these effects.

Spherical aberrations prevent the light rays, from a point source, entering the lens from focusing at a common point and thus the image spreads into a blurred circle. The effect is particularly noticeable when using transmissive optics with a small radius of curvature on their back face; hence short focal length, low f number, plano-convex lenses are particularly vulnerable. Spherical aberration can be reduced by computer programmed, diamond turning of specially configured aspheric lenses to suit the laser beam diameter and beam characteristics. Such optics are expensive and generally make too narrow welds, but are excellent for laser cutting.

Transmissive optic diameter and edge thickness are best specified by an optics manufacturer. To make their recommendation they need to know the following details: optic material required; the laser beam mode and beam divergence angle (consult laser manufacturer or handbook); the laser beam diameter at the proposed laser gun position; the laser light wave length; the laser power to be focused and the focal length required.

### Handling and mounting transmissive focusing optics

Borosilicate crown glass, KCl and ZnSe are all delicate materials and to be effective require high surface finishes. A small scratch, dig or fingerprint can ruin their optical performance and therefore they should always be handled with considerable care. Ideally, new finger cots or, polyethylene gloves should be worn to avoid finger prints. It is also advisable to avoid handling the central region of the lens to eliminate the risk of scratching even if wearing finger protection. For handling ZnSe it is recommended that polyethylene gloves are always worn, especially when removing a cracked or damaged optic, because ZnSe is poisonous and small particles entering cuts or grazes can be extremely dangerous.

When mounting transmissive optics, they should not be mechanically stressed by their clamping system, otherwise, because of their delicate nature, they will be prone to premature failure due to cracking. The quality commercial optic mounts, which are designed to screw directly into a laser gun, Fig. 11.2, usually sandwich the lens between two

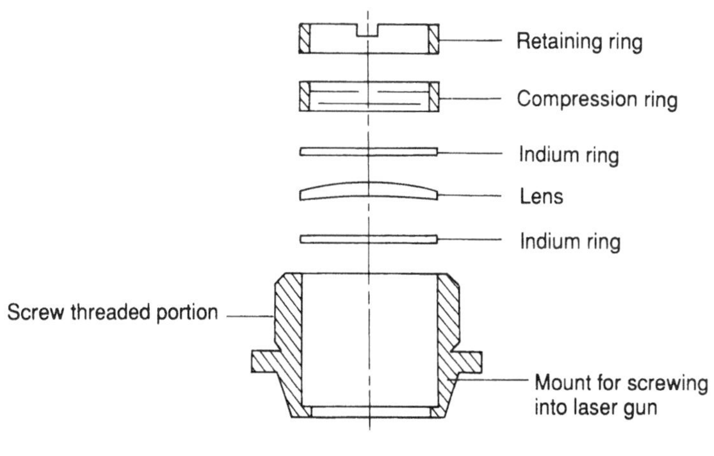


**11.2** Construction of a typical commercial focusing optic mount and component parts.

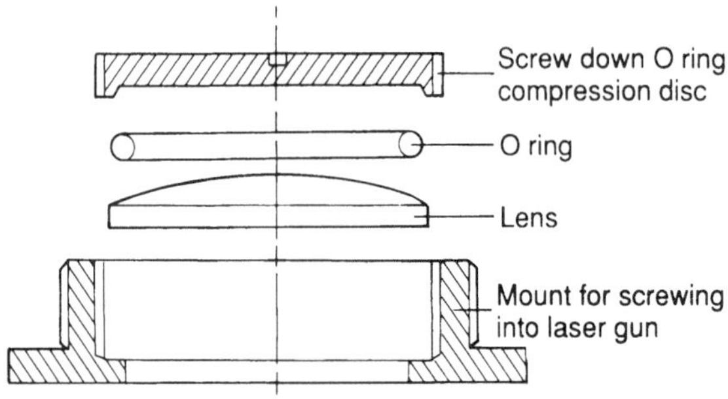


**11.3** A simple optic clamping system which can be made in-house.

indium rings and employ a special compression ring to equalise the clamping pressure. The indium metal contact also assists, by acting as a heat sink, to keep the optic cool and thus limit beam absorption. This type of mount is not, however, absolutely essential. For most $CO_2$ laser welding applications a simple clamping system, as shown in Fig. 11.3, where an 'O' ring under light compression pushes the lens front face against a hard smooth surface is quite acceptable, that is provided the lens cannot be moved under the pressure of the shield gas where a co-axial nozzle is used. Whichever type of mount is used, it is most important that it is positioned so that the lens is square to the incident laser beam axis and that the lens optical axis is also as concentric as is practically possible with that of the beam.

## Reflective focusing optics

Reflective optics are not used to focus Nd:YAG laser beams because standard glass lenses which can be produced almost aberration-free, are very compatible with the small diameter beams from Nd:YAG lasers. Also such lenses when used back-to-back easily make the integral beam collimator/focusing heads which are usually necessary when operating Nd:YAG lasers.

For focusing $CO_2$ laser light, plain copper (oxygen-free high conductivity), gold coated copper and molybdenum focusing mirrors are used. Copper, which can be used with highly polished or diamond turned surface finishes, has a reflectance of approximately 99%. It is, however, easily scratched and to combat this some manufacturers apply a protective dielectric coating. It is also prone to corrosion pitting. Gold coatings combat pitting and also marginally improve reflectivity.

Molybdenum has a reflectance of approximately 98%, but is considerably more durable than the other materials mentioned. Weld spatter freezes immediately it hits the surface of molybdenum and can usually be easily chipped off with a nylon, or similar soft material, blade without causing any damage to the optical surface. Moreover, it will withstand many cleaning operations before reworking by the optic manufacturer becomes necessary; the cleaning of optics is described shortly. Although molybdenum focusing mirrors are more expensive than those based on copper, the considerably increased service life outweighs the extra cost.

Reflective focusing optics are generally in the form of a conventional concave mirror or a parabolic mirror. For the concave mirror, the angle of incidence of the beam on to the mirror has to be as near normal as possible, ideally within 10° of the optical axis or better, otherwise the focus spot will be too large and have a pear shaped footprint. To achieve the required angle of incidence on the concave mirror, a fold mirror has to be positioned below it and to one side, Fig. 11.4, which makes the laser gun rather bulky and difficult to use and, if necessary, manipulate in confined spaces, due to the gun body being too close to the work surface. A parabolic mirror, Fig. 11.5, does not necessarily need such a pre-focus mirror position, since its parabola can be configured to accept a laser beam at 90° to the focus axis. The advantage of this is realised when multi-axis movement of the laser gun is required (see Chapter 10, Fig. 10.5) as it allows the parabolic focusing mirror to be rotated through 360° about its incident beam axis.

The main advantage of the parabolic mirror over the concave mirror is of course that the parabolic mirror produces a smaller and circular focus spot and is therefore more efficient in terms of power density level and control. Nevertheless, the parabolic mirror in the form so far described (Fig. 11.5) has two major disadvantages. Firstly, it is very expensive to produce since the parabola section has to be machined, by diamond turning, from a considerably larger piece of material than the final mirror produced, and secondly, the alignment of the laser beam incident on the parabolic mirror is critical and if not correct the focusing advantage over a concave mirror is easily lost. Fortunately, parabolic focusing mirrors and their pre-focus beam fold mirror can now be purchased pre-aligned in a laser gun, which by following the manufacturer's instructions can be accurately aligned with the incoming laser beam.

In instances where the parabolic mirror is not required to be rotated

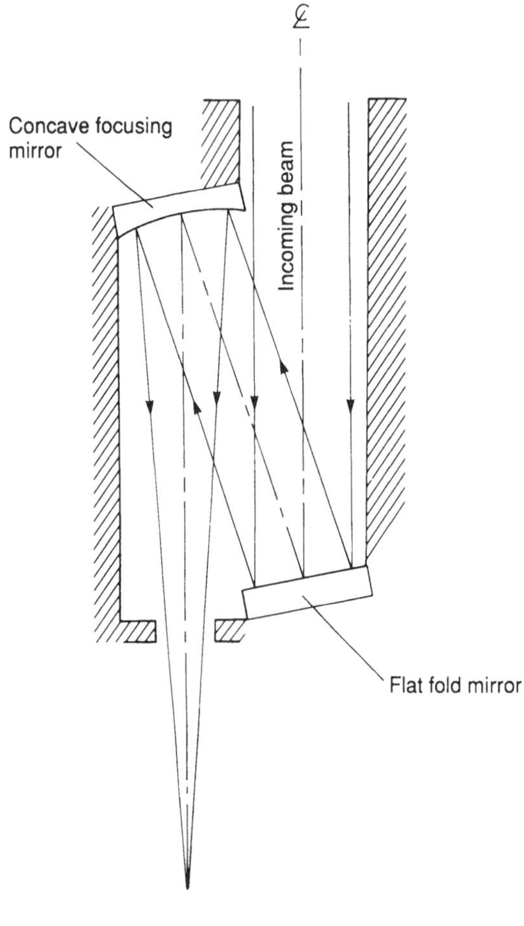


11.4 The necessary mirror arrangement when using a conventional concave mirror for focusing a laser beam. The fold mirror must be positioned so that the centreline of the beam reflected from it strikes the focusing mirror optical axis at an acute angle (ideally less than 10°).

about its incident beam axis, a cheaper more easily produced parabola from a smaller piece of material can be made if the incident beam is more acute to the angle of reflection (as in Fig. 11.4). Although this makes a more bulky laser gun, increased stand-off distance between the laser gun body and the work surface can be achieved by using a beam expansion mirror, Fig. 11.6, placed before the focusing mirror. When employing this technique, one must not forget to match as near as possible the beam diameter incident on the focusing mirror with the

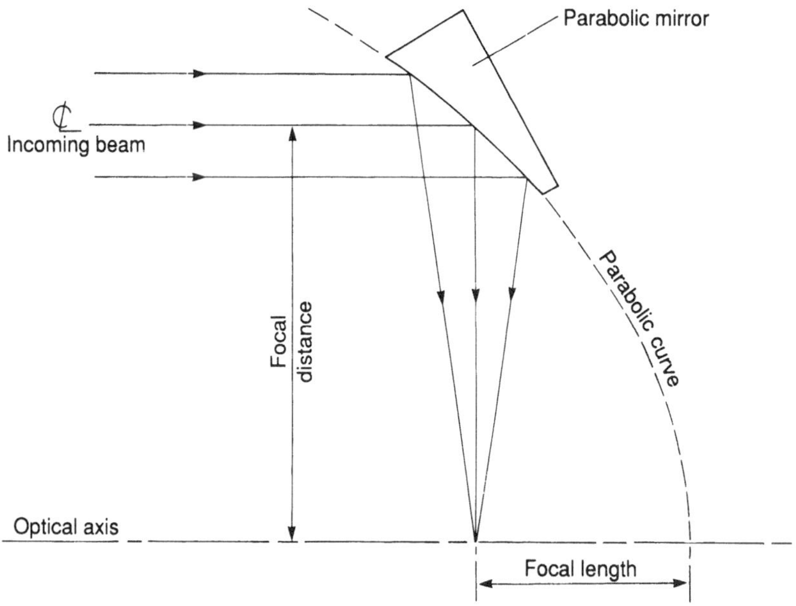


11.5 The concept of a 90° off-axis parabolic focusing optic.

necessary focal length for achieving the required focus f number. It should also be noted that the increased laser gun stand-off, afforded by using a pre-focus beam expander, also increases the focusing optic life by taking it further from the weld zone. It also enables space for a transverse gas jet (to be described shortly) for providing further protection to the focusing optic from weld metal vapour and spatter.

## Which focusing optic – reflective or transmissive?

As described earlier, transmissive optics are without doubt the most practical at present for Nd:YAG laser welding; however, for $CO_2$ laser welding one is confronted with the choice of transmissive or reflective focusing optics and therefore it is necessary to consider carefully the advantages and disadvantages of each.

Transmissive focusing optics are more easy to set up and enable more simple laser guns to be used and consequently they generally give tighter focus spots than reflective optics. They will comfortably handle laser powers up to 3 kW, but above 5 kW they are prone to thermal distortion, especially when transmitting high power $TEM_{00}$ mode

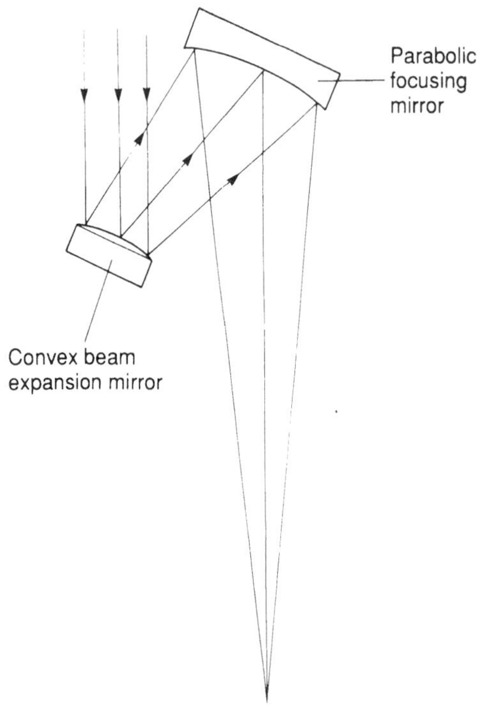


**11.6** A method of increasing the optic and laser gun stand-off distance from the workpiece, by using a convex beam expansion mirror placed before a parabolic focusing mirror.

beams where the energy density is highest at the centre, which will cause a variation in focal length and hence a change of focus position. Transmissive focusing optics are readily available in a range of diameters and focal lengths and most manufacturers will produce special focal lengths without a significant increase in cost. Transmissive optics are sensitive to contamination and if slightly contaminated will quickly absorb laser power, lose their focusing ability and fail through cracking. Finally, KCl cannot be used where there is high humidity and is therefore generally confined to welding research and development.

Reflective focusing optics are more difficult to set up to focus the laser beam as well as high quality transmissive optics and to produce an efficient focus spot they need to be incorporated in precision made laser gun housings to ensure their correct alignment with the incoming laser beam and fold mirror train. They do not suffer from thermal distortion because they are uniformly water cooled (the cooling water temperature should always be near the ambient, otherwise condensation, especially in hot humid weather, will form on the mirror face and scatter the laser light elsewhere). They cannot suffer catastrophic failure through cracking and therefore their performance is only downgraded by contamination or changes in the laser beam. Reflective optics can be reworked when their surfaces are damaged.

Molybdenum mirror faces can be cleaned many times and are perhaps the most resistant to weld spatter damage.

From the foregoing it can be gathered that reflective focusing optics, especially molybdenum, are generally more practical for production welding where equipment reliability is paramount. Moreover, in such circumstances they are essential for laser powers much above 3 kW. Nevertheless, when working at lower powers, transmissive optics enable simple laser guns to be used, especially where a capstan lens system is required to change focal lengths quickly for different processing operations, e.g. as on a mixed welding and cutting station. Incidentally, transmissive optics are generally far superior to reflective optics for high quality laser cutting work.

## Cleaning laser optics and in-process protection

Laser focusing and beam transmission optics are all prone to thermal distortion and the associated loss of performance and damage if they become contaminated. This is because contamination absorbs laser light which generates heat. Since hot surfaces absorb energy more efficiently from laser light a thermal runaway effect quickly occurs, which causes thermal distortion and other more catastrophic forms of optic damage.

Dust, oil mist, water vapour, metal vapour, weld spatter and fingerprints will all cause absorption if allowed to form on a laser processing optic. To prolong optic life, steps need to be taken to keep potential contaminants away and to provide regular cleaning at realistic intervals. The period between cleaning operations will depend on the working environment, the availability of some form of in-process protection system and the optic attitude; an optical face which is pointing vertically upward will be more prone to contamination by settling dust than one which is pointing downwards or horizontally. Ideally, optics should be visually checked at the beginning of each shift for signs of unforeseen contamination and damage. The cleaning can be established practically in the early days of operating the equipment. This is best done by keeping records of when the optics show signs of contamination and when related changes in welding performance first occur and then reducing the recorded time span to a safe period before changes can have any effect on welding performance.

### Cleaning procedure

It is extremely difficult to clean the surfaces of KCl optics, other than removing dust and other free standing particulates with the aid of an air bulb or an aerosol clean air spray. Workshop airlines should not be used to blow particles off optics, because they usually contain fine particles from within the airline pipework (especially if the pipework is old and rusty) and often significant amounts of oil and water.

In comparison to KCl, glass, ZnSe and gold coated and plain copper reflective optic materials can withstand careful surface contact cleaning operations to remove lightly adhered contamination and thus prevent it building up to an unacceptable level. The procedure is as follows:

1 If possible remove the optic (this may not be practical in a multi-mirror $CO_2$ laser beam transmission train) and blow off any loose particles with an air bulb, and particularly with a transmissive optic, place on a clean soft cotton cloth. Remember when handling ZnSe to wear polyethylene gloves or rubber finger cots (ZnSe is a poisonous substance) and always hold all optics by their edges.
2 Using a clean surgical cotton ball, soaked in acetone, gently wipe the surface clean.
3 To lessen the risk of acetone evaporation stains and contamination streaks on the optic surface, flood the optic surface with acetone. Then holding a lens tissue by the corners on one side, lay it on the optic and drag it slowly across the optic surface at a rate which will absorb acetone and enable any remaining to evaporate against the trailing edge of the lens tissue.

When cleaning the second face of a transmissive optic, action 3) can be difficult to achieve without re-wetting the first face. Therefore, for cleaning the second face, the optic is best carefully held on edge by an assistant, or in a suitable clamp.

Sometimes, badly contaminated optics can be cleaned by first using an optical polishing compound (these are best recommended by the manufacturer of the optic in question) before following the cleaning sequences described above. The compound should be applied with a soft, clean surgical cotton ball and using a light pressure. Also, keep the polishing action to the minimum necessary for removing the contamination. Excessive polishing can remove special surface coatings often applied to optics to enhance their performance. If coating damage does occur (on ZnSe a patch with a different surface colour appears), remember that it may not be necessary to discard the optic, because it may be able to be reworked, by the manufacturer, to as-new condition and at a cost considerably cheaper than that of a new optic; the major cost of a ZnSe optic is in the raw material.

## In-process protection

The focusing optics of Nd:YAG lasers are easily protected by simple and relatively cheap glass cover slides, incorporated just below the lens in the focusing gun. The slides are specially produced with the necessary optical quality for the application and therefore unspecified glass slides should not be used as they could disrupt focusing of the beam. Cover slides will gradually become contaminated and therefore they should be regularly changed before the level of contamination influences weld quality.

The use of cover slides is generally impractical for protecting $CO_2$ laser beam focusing optics because of the high costs of the special window materials necessary to transmit the wavelength of $CO_2$ laser light. Therefore, for protecting $CO_2$ focusing optics a simple gas cross jet manifold is recommended. Such a device in relation to the focusing optic and work gas shielding arrangement was shown earlier in Chapter 5, Fig. 5.15. Basically it comprises a gas fed manifold with a line or several lines, of small holes which blow a laminar gas stream in a direction transverse to the laser beam. Thus hot metal vapour, spatter and soot drifting up towards the optic from the weld zone is blown off course. For the protection gas, bottled compressed air or nitrogen gas can be used, the latter being preferable if the cross jet is relatively close to the work; as the oxygen in the compressed air could affect the weld pool. Optic life will be enhanced the further the optic and cross jet are from the work. This is not always possible if short focal length optics are used. Nevertheless, in such cases larger stand-offs can often be achieved by using a beam expander (see Fig. 11.6) in front of the focusing optic, thus enabling the use of longer focal length whilst preserving the required f number.

## Laser beam transmission systems

The previous chapter described how the laser beam can be manipulated about the workpiece and also transmitted to numerous workstations. Here, a general introduction to the concepts of Nd:YAG and $CO_2$ laser beam transmission systems is given, together with practical operating advice, which may be beneficial to the potential user. The information presented is not aimed at the development of in-house equipment, because extensive optical equipment know-how is required and consequently such development work is best left to experts in this field.

### Fibre optic (for Nd:YAG lasers)

As shown in Chapter 10, fibre optic beam delivery systems are by far the most simple and versatile. Their concept in relation to the laser and focusing head (laser gun) is shown schematically in Fig. 11.7. The optical fibre material is $SiO_2$ (quartz) and is generally less than 1 mm diameter. For maximum beam transmission efficiency there is an optimum relationship between the laser beam output diameter, its divergence angle and the optical fibre diameter, and, consequently, the lens which focuses the beam into the fibre. The relationship is referred to as the numerical aperture and this is best specified, together with the necessary optical equipment, by the laser manufacturer.

For maximum beam transmission efficiency the fibre ends have to be highly polished and perfectly square and concentric with the optical axis of the lenses positioned at each end of the fibre. Moreover, the focus position of the incoming beam has to be precise in relation to the end of the fibre. These adjustments require some skill, and expert training by the laser supplier is recommended. Beam transmission efficiency is also impaired if the fibre is bent too tightly. A 0.5 mm diameter $SiO_2$ fibre has an allowable bending radius of approximately 100 mm, before efficiency is impaired, whereas for a 1 mm diameter fibre the safe radius is at least twice as large. Generally the total loss of laser power for an Nd:YAG laser and fibre optic system is no more than 10–15%.

The optical fibre assemblies used to transmit laser welding powers are purpose-made and quite different to those used in electronics. Those used for welding are protected by substantial and robust sheathing, which includes a flexible steel tube and nylon jacket. Although these adequately protect the optical fibre, their main function is to resist

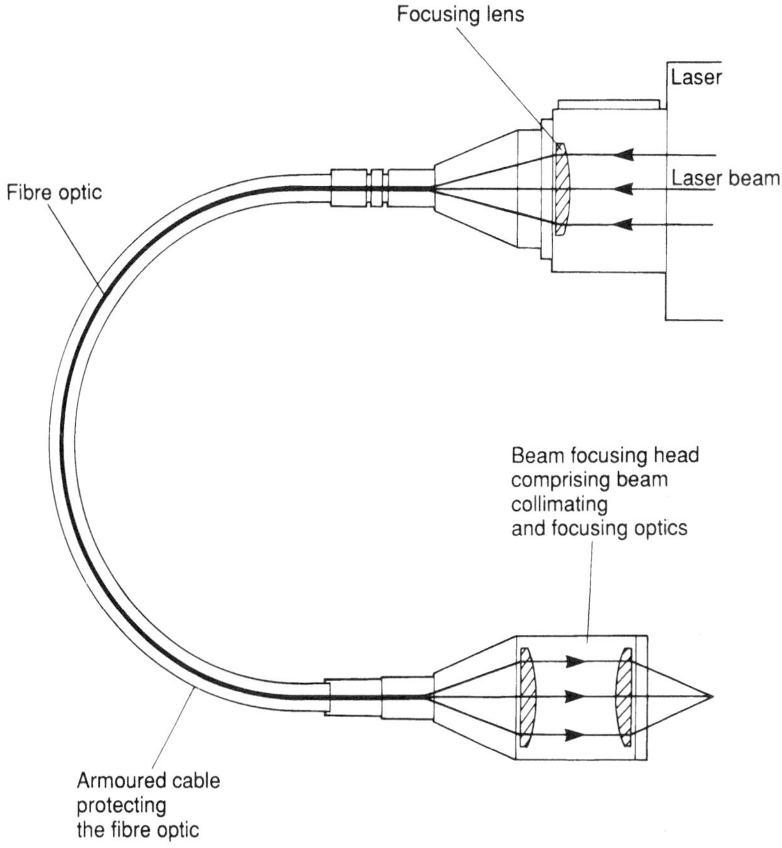


**11.7** A fibre optic beam delivery system.

accidental industrial damage which could cause fracture and allow laser light to escape. As an extra precaution against the latter, the sheathing can incorporate an electrical circuit which automatically switches off the laser if the sheathing is damaged from the inside or out. Despite these precautions, fibre optic beam transmission cables should always be routed with care and regularly inspected for damage. Where fibre optic cables are used in conjunction with robots, extra special care must be taken to ensure the robot operating program will not entangle, trap, or bend the cable beyond its allowable bending radius.

### Reflective optic (for $CO_2$ lasers)

The basic concept of a reflective optic beam transmission system can be grasped from the over-simplified form which is shown schematically in Chapter 10, Fig. 10.5. What the illustration does not convey is that, from the laser output window, each mirror in the train to the laser gun outlet, needs to be water cooled and precision aligned with its neighbour. Furthermore, between each mirror a protected beam path is necessary, both for personnel safety and protection of the laser beam itself. Each mirror needs to be mounted on a vibration-free and thermally stable support. Experience has shown that some commercial equipments can meet these requirements, whilst others have fallen far short. From the user's standpoint, the most reliable beam transmission system will be that which meets the above requirements, uses the least mirrors and keeps the beam transmission distance to the bare minimum.

Water cooling the mirrors presents few problems if correctly carried out and the system operation is regularly monitored. Never connect cooling water pipes in series to a mirror train, otherwise the mirrors towards the end of the train will receive hot water and the cooling effect will be lost. Some mirrors have screw-on cooling water connections and care must be taken not to disturb the mirror alignment when fitting these. Simple snap-on connectors reduce the risk of such mirror disturbance. To ensure an adequate and consistent flow of water through the mirror train, water flow gauges fitted with warning sensors are recommended.

Accurate alignment of the laser beam through any form of mirror train is vital, especially where a mirror is required to be rotated about its incident beam axis, as in the case of an articulated wrist as used with a moving laser gun. The importance of mirror-to-mirror alignment can be appreciated if one considers two mirrors 10 m apart, which are aligned centre-to-centre. If the transmitting mirror is then very slightly moved to deflect the beam as little as 0.5 milliradian, the beam will move 5 mm from the centre of the receiving mirror; this example also emphasises the need for vibration free and thermally stable mirror supports. Alignment of a mirror train after a complete change of mirrors has to start at the first mirror after the laser output window, or, if just one mirror has been changed, at the first mirror after the changed mirror.

The fold mirrors, which transmit the laser beam to the laser gun assembly, generally have two forms of adjustment. The first is an adjustment for positioning the mirror mount itself, Fig. 11.8, to receive the incoming beam correctly and the second is to provide precision tilt and twist for accurately deflecting the beam to the next station. The first adjustment is normally only necessary when the overall system is installed, or if the laser has had to be moved or replaced.

Unfortunately the alignment of mirror trains can only rarely be conducted with the HeNe pointing laser (Chapter 1, Fig. 1.11), which is folded in co-axially with the $CO_2$ laser output beam. This is because over distances of several metres acceptable co-axial alignment is usually lost. Under these circumstances a reliable method of aligning a mirror train is to make acrylic (e.g. Perspex) targets, and take burn prints using the laser beam. The targets, which need to be at least 6 mm thick and substantially larger than the expected incident beam diameter, should have cross hairs or concentric rings machined on their surface and if possible they should be fitted in the aperture normally occupied

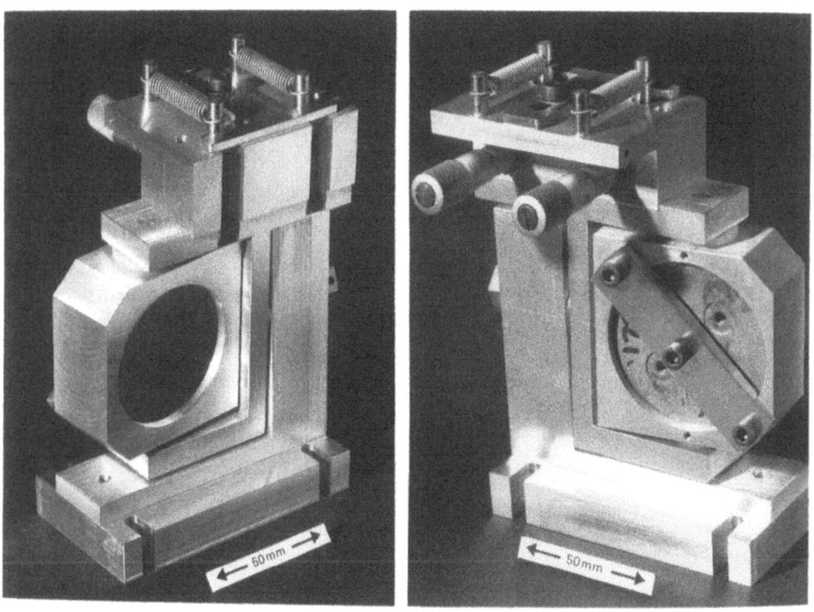


**11.8** Front and rear views of a laser beam fold mirror assembly. The slots in the base are for initially mounting and positioning the assembly to receive the laser beam. The two micrometer thimble adjusters at the back provide the precision tilt and twist for accurately deflecting the laser beam to the next mirror or station.

by the mirror. If the latter is not practical, which usually tends to be the case, the target should be positioned in a retort stand or similar clamping device close to the mirror and with its centre aligned as close as visually possible to that of the mirror. Also a carbon block should be placed between the target and the mirror in case the beam is accidentally allowed to burn through the target. Before proceeding with the alignment procedure the laser resonator must be in a correctly tuned state, i.e. with its fully reflecting mirror and output windows (Chapter 1, Fig. 1.11) in their optimum positions for maximum laser power. This requirement is necessary because the fully reflecting mirror and the output window positions collectively determine the direction the laser beam points as it leaves the laser. With the laser pointing direction established, the procedure for aligning the mirror train is as follows:

1 Place an acrylic target in the position of the first fold mirror. Position an airline to give an air blast across the target surface during beam printing as illustrated in Fig. 11.9. This is necessary to remove smoke and prevent the acrylic from catching fire.
2 Using a low laser power, open the laser shutter for 2 or 3 seconds and produce a print of the beam profile on the acrylic target. Depending on the laser power, some experimentation with the print time may be necessary to produce a clear burn profile. *Caution*, prior to making the beam print ensure all personnel are behind safety screens and that the exposed laser beam between the laser and the fold mirror position cannot strike or reflect off any objects.
3 Examine the beam print position in relation to the target lines. On the first mirror in the train from the laser, the print should be on target (subject to the correct initial installation) and if not the mirror mount should be carefully adjusted. Several further burn prints may be necessary to perfect this alignment.
4 Once the print is on target, align the HeNe laser also on the target. Then remove the target and replace the mirror or remove the carbon block as the case may be.
5 The HeNe laser should now be on the next mirror in the train and thus confirm that the beam reflected from the first mirror is approximately on target. To see the HeNe beam gently place a lens tissue on the second mirror face. If the HeNe beam is not exactly on the centre of the mirror, but well in from the edge, do not immediately adjust the tilt and twist of the previous mirror; the HeNe beam may not be concentric with the laser beam at this point. If the

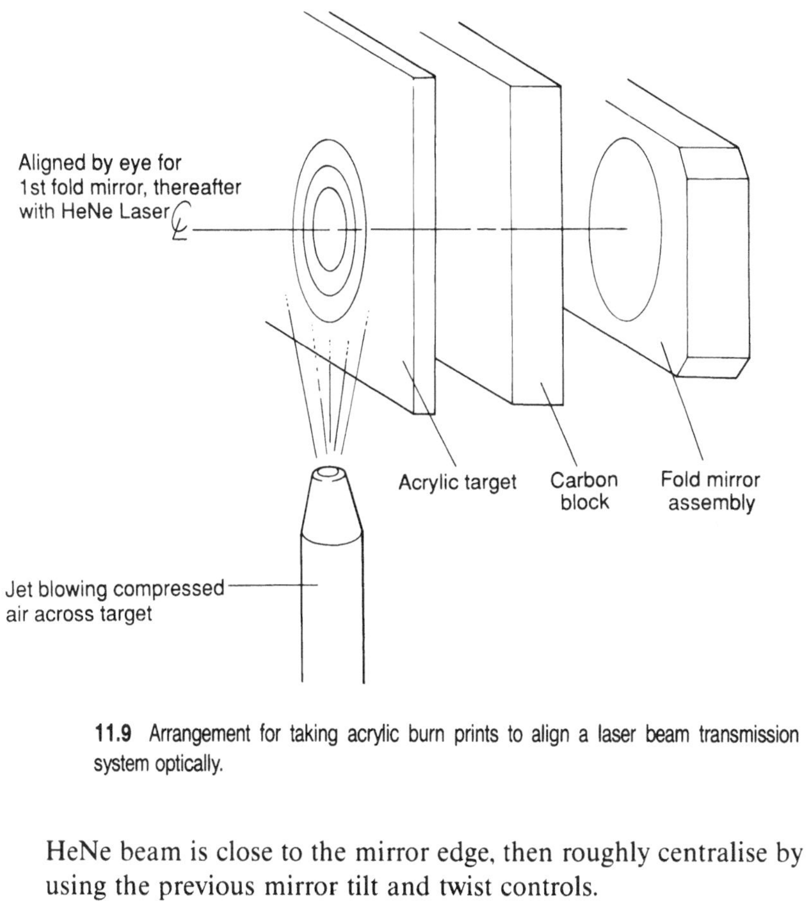


**11.9** Arrangement for taking acrylic burn prints to align a laser beam transmission system optically.

HeNe beam is close to the mirror edge, then roughly centralise by using the previous mirror tilt and twist controls.

6 Place an acrylic target in the position of the second fold mirror and repeat the laser beam printing procedure. If the print is off target adjust the tilt and twist control of the previous mirror and take further beam prints until the correct alignment is achieved.

7 From now on repeat actions 4) through to 6) and work through the remaining fold mirrors in the train.

Aligning mirror transmission trains is not easy and success comes only with extreme patience and attention to detail.

In general use, laser beams must never be transmitted from one mirror to another without a protective cover. The most important reason for this is safety, which is discussed in the next chapter. However, the laser beam itself needs protecting otherwise an effect referred to as thermal blooming can occur, which disrupts the laser beam and reduces power density and focusability. Thermal blooming occurs as a result of airborne particles and vapours obstructing the laser beam path, causing beam energy to be absorbed and light waves to be reflected from their normal path. To avoid these problems, laser beams are normally transmitted down tubes made of aluminium alloy (e.g. international designation 6082). Aluminium alloy is used as it reflects the beam, rather than absorbs it which would cause tube burn through in the event of the beam clipping the tube wall through accidental mirror movement.

In an environment of oil, water or cleaning agent vapour, beam transmission tubes alone will not always stop thermal blooming. In such cases it may be necessary to purge the transmission tubes with dry nitrogen, or insert an air mover, with an incoming air moisture and particulate filter, in the system.

On large beam transmission systems, maintaining optical alignment is usually the major maintenance requirement and the level necessary is directly related to the quality of the equipment and the pointing stability of the laser. The stability of the mirror mounts and their support structures are perhaps the most important factors. The mirror train alignment being correct, the other most common causes of poor transmission of the beam power and hence weld energy density are usually mirror contamination, poor mirror cooling, thermal blooming during beam transmission and thermal distortion of the transmissive optic used for the laser output window (some lasers use aero-windows instead of a solid transmissive optic and these are described shortly).

$CO_2$ laser output windows of the transmissive optic type have limited operational lives. This is because they are prone to contamination and thus thermal distortion by the reactive environment in the laser cavity causing absorbing deposits to build up gradually on the internal face. Such contamination, however, also produces a drop in laser output power for a given resonator excitation current and therefore a drop in laser performance, simultaneously with a change in burn print shape, or laser beam analyser recording, is a good indicator that the beam transmission train may not be at fault if the beam power and profile is found to have deteriorated at the end of the transmission train.

### Aero-windows

Laser output windows of the transmissive optic type cannot easily and continually handle high laser powers in the order of 10 kW and above without suffering thermal lensing, which varies the propagation of the transmitted beam. Consequently, aero-windows working in conjunction with reflective optics have been developed to overcome this problem. Basically an aero-window is a short small bore, typically 6 mm diameter tube, with an internal pressure equal to that in the resonator, and which allows the laser beam out of the resonator without letting in the local atmosphere, which would destroy the lasing action. The low pressure required in the small tube is achieved by using a high velocity air stream passing across the end of the tube (Benoulli's principle). The arrangement in relation to the necessary reflective optics is shown schematically in Fig. 11.10. The portion of the laser beam which is required to leave the resonator is folded on to a focusing mirror which brings the beam to a small waist diameter (typically 3 mm) in the small aero-window tube, where the required pressure is controlled by the high velocity cross jet of air. On leaving the window the now expanding laser beam is re-collimated to a near parallel beam by a pair of specially configured mirrors.

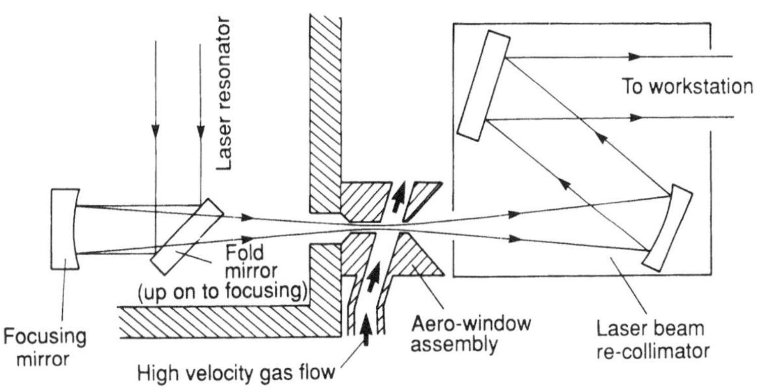


11.10 An aero-window and associated laser beam control optics.

## References

1 Greening D E: 'Advances in optics for $CO_2$ lasers'. *Engineer Lasers*, Jan 1990.
2 Sherman G H et al: 'Latest optics for industrial $CO_2$ lasers'. Int Conf Laser Material Processing, Science and Applications. Osaka, Japan, 21-23 May 1987.
3 Herrit G L and Scatena D J: 'Choose the right mirror for industrial $CO_2$ lasers'. *Laser Focus World*, July 1991.
4 Reedy H E and Herrit G L: 'Comparison of GaAs and ZnSe for high power $CO_2$ laser optics'. SPIE Volume 1020.
5 Scatena D J and Herrit G L: 'How to avoid contamination problems in $CO_2$ laser optics'. *Laser Focus World*, December 1990.

Chapter 12

# Laser safety

Laser welding systems are very safe provided they are operated and maintained in accordance with safety standards, legal requirements and manufacturers' operation and maintenance procedures. To ensure these requirements are met, a Laser Safety Officer should be appointed who has knowledge and training in the necessary subjects. Owing to the infancy of laser welding technology, such a person usually has to be specially trained and most laser manufacturers will provide this service; a point worth establishing when acquiring a laser or laser system.

Legal and safety requirements for safe working with lasers do vary from country to country. In Europe, the EEC and EFTA member countries have agreed on a safety standard, proposed by the European Committee for Electrotechnical Standardization (CENELEC), for non-ionizing radiation safety of laser products; CENELEC Harmonization Document HD 482 S1, which is in turn based on IEC Publication 825:1984. In the United Kingdom the implementation of CENELEC 482 S1 is presented by BS 7192:1989. In the United States guidelines on the safe use of lasers are presented in ANSI Z1316.1.

> This chapter provides some general safety information, precautions and recommended publications dedicated to the safe use of laser equipment. Those using, or seriously considering using, laser welding are strongly advised to acquaint themselves with their relevant national requirements. The information presented here is intended to identify the most important safety aspects associated with lasers used for welding and provide a guide for their safe use. It is by no means a complete list of requirements and the author and publisher accept no responsibility for incidents arising out of sole reliance on this guide.

The hazards presented when operating a laser for welding (or any other form of high power laser processing) fall into three general groups: laser beam, electrical, and chemical and fume.

## Laser beam hazards

An unfocused laser beam, escaping from the beam transmission path through the omission of beam transmission components, is capable of being transmitted over a distance of several hundred metres in air before its natural divergence will cause it to expand to a safe diameter where the power density is too low to cause damage. For welding, the laser beam is focused to a point where the power density is very high and if incident on the skin very deep burns, or even serious burn holes, would occur. Beyond the focus point the beam normally expands more rapidly than the unfocused beam, generally reaching a safe diameter after just a few metres. The precise distance depends on the focus f number; the lower the number the greater the rate of beam expansion. A hazard may arise due to the focused beam passing through or reflecting off the workpiece. Such hazards can arise particularly when the laser beam strikes a bright metal surface at an angle in the order of 20° or greater from the normal position (at right angles to the work surface).

Unguarded laser beams or escaping reflected laser light from the weld zone can cause serious eye damage as well as skin burns to anyone caught in the light path. One must remember that the light from $CO_2$ and Nd:YAG lasers is invisible to the eye and even if it could be seen, because it travels at 186 000 miles per second, it would almost instantaneously strike anyone unfortunate enough to be in its path.

The light wave length from an Nd:YAG laser (1.06 μm) is particularly dangerous to the eye, because the lens of the eye can focus this wave length to a very small spot on the retina and cause blindspots. Apparently, the retina does not register pain, and damage in the form of blindspots is not always realised immediately. The wave length from a $CO_2$ laser (10.6 μm) is not transmitted by the lens of the eye and in this case damage occurs at the front of the eye, with burns on the cornea and cataract formation. Regular eye tests for laser personnel are recommended so that records are kept, in case of eye damage.

The potential risk of burn damage to the body should never be underestimated, as gruesome as it may sound, an unfocused high power laser beam of several millimetres diameter if incident on the body could maim one for life, (such accidents are perhaps most likely to occur where the laser beam is manipulated about the workpiece by a robotic system and the casualty has invaded the safety enclosure and overidden safety devices).

Apart from damage to the person, escaping laser light can start fires and easily melt pipework and cable covers and thus lead to other unwanted dangerous situations by affecting the safe operation of other plant. One should not forget an unfocused beam from a multi-kilowatt laser, if given time, will easily burn through fire bricks and steel plates.

The HeNe pointing lasers which are folded into the beam paths of welding lasers present no significant danger if their maximum power rating is less than 0.5 mW. However, some systems have HeNe lasers slightly more powerful, typically 0.8 mW. Direct viewing of beams above 0.5 mW must be avoided. Specular reflections must also be avoided and any targets used to locate the beam must also scatter the beam to avoid a focused reflection.

## Electrical hazards

All laser power supplies generate high voltages. Furthermore, high power lasers incorporate capacitors and various potentially dangerous components which can maintain their stored energy when the system is switched off. Moreover, some $CO_2$ lasers have RF excitation, which must be screened when operated to protect personnel from this form of radiation. All in all lasers are extremely hazardous from an electrical standpoint. In fact, it has been found in the USA and Europe that laser related accidents of an electrical origin greatly exceed those involving eye and skin injuries. Several people have been electrocuted whilst troubleshooting or servicing laser equipment. Many accidents involved people working alone on high voltage circuits. This highlights the importance of adequate training, working in pairs and working to correct procedures. The HSE Electricity at Work Act advises only trained personnel to work on electrical equipment.

## Chemical and fume hazards

The spent lasing gases slowly exhausted from $CO_2$ lasers contain carbon monoxide, care must therefore be taken if a laser is operated in a poorly ventilated area. Ideally, a gas exhaust pipe should be passed outside the building to a point well clear of any air intakes and occupied areas.

Laser output windows and transmissive focusing optics made from gallium arsenide (windows) and zinc selenide (windows and optics) are potential toxic hazards if damaged. The hazard arises if these materials are overheated, which causes them to decompose and oxidise, resulting in the formation of toxic oxides of arsenic and selenium. These appear as powdery deposits, which settle close to the point of thermal damage, and have to be carefully dealt with; the suggested method is described below.

When fusion welding metals, airborne contaminants are released in the form of metallic vapour and dust, metallic oxide fume, and chemical and gaseous vapours. All of these can be hazardous to health especially if the metal contains dangerous elements such as cadmium and beryllium. Accordingly, all fume should be regarded as potentially harmful and steps taken to prevent inhalation.

## Practical precautions

### Laser beam precautions

- All work relating to laser beam transmission and focusing systems such as operation, installation, modification and maintenance, must be conducted by trained personnel.
- The complete laser beam transmission path, from the point where the beam leaves the laser to where it exits the laser gun, should be enclosed. All access panels to the beam transmission path should have remote electrical interlock connectors, which automatically revert to open circuit and shut down the laser beam transmission when a panel is removed. When re-aligning the mirror train of a $CO_2$ laser it will be necessary to override these safety switches. In this case, a safe working procedure must be established and agreed with the Laser Safety Officer, adhered to, and the safety interlock system re-instated immediately the mirror alignment is complete.
- The laser gun, whether stationary or moving must be operated inside a safety enclosure. All access doors should have interlocks which terminate beam transmission if the doors are opened, as a double safety measure, the same key for operating the laser can also be used to open the work enclosure doors, thus making it necessary to switch off the laser to enter the enclosure.

- Laser beam transmission tubes and fibre optic cables should be positioned so that they cannot be struck and damaged by overhead cranes, vehicles and any general workshop practices.
- The safety enclosure walls must be made of a material which will absorb the laser beam and not catch fire (steel or brick are recommended). They must also, however, be at a distance from the stray laser light source, where the laser power density will be too low to promote melting and possible burn through. In the case of stationary laser guns these two requirements do not present a major problem, since the laser gun and gas shielding device will considerably limit the area through which any reflected light can escape. Light reflects at an equal and opposite angle to the angle of incidence and therefore, if the gun is positioned normal to the work surface, any light reflected will generally reflect back on to the gun or gas shield and be absorbed by these. (In twelve years of operating laser guns transmitting powers from less than 1 kW up to 8.7 kW in the down hand position and at a distance of 1.5 m from the enclosure wall, the author has never seen any signs of burn marks through stray reflections. Nevertheless, becoming complacent and taking risks are out of the question as there is a first time for every kind of accident.) The greatest risk comes from laser guns which can be manipulated about the work and hence point in any direction. With these the gun can move away from the work, through accident or intent and point directly at the enclosure wall. In this case all the laser beam, and not just a reflection of a part, will fall incident on the enclosure wall. Fortunately the beam will be diverging. At a distance from the laser gun equal to the focal length of the focusing optic the beam diameter will equal the beam diameter incident on the focusing optic and will still have a dangerous power density (typically 10 W/mm$^2$). However, a further five focal lengths away the beam will be approximately six times larger in diameter and the power density will have dropped by thirty-six times and to a less dangerous level. However, depending on the initial laser power the level could still cause skin burns and eye damage. (To avoid skin burn and eye damage the power density from a $CO_2$ laser beam must be below 0.1 W/mm$^2$.)

  What all this means is that, where a gun can be manipulated, the enclosure wall should ideally be a sufficient distance from the gun's maximum range of operation so that the power density of the diverging beam as it strikes the wall cannot cause, under any circumstances, sufficient heating to achieve serious melting and burn through of the enclosure.

- Viewing windows for Nd:YAG laser enclosures are made of a special material usually available from the laser manufacturer and which is quite expensive. Glass and acrylic sheet are transparent to the 1.06 μm wavelength of Nd:YAG laser light and therefore must not be used unless a special absorbing film coating has been added. Owing to the high risk of serious eye damage from an Nd:YAG laser, a closed-circuit television system is recommended, instead of a window, for viewing the welding operation; with the correct camera and filters close-up observations can be made in absolute safety.

- Viewing windows for $CO_2$ laser enclosures are best made from an acrylic material such as Perspex. If an errant laser beam of sufficient power density hits this material it quickly gives off black smoke and would alert the operator and other personnel (if the power density is high enough to burn such a window, the window is too close to the point of initial beam focus, or the errant beam is being refocused; the latter is most unlikely when operating laser guns in the downhand position). A good and safe window system for a $CO_2$ laser enclosure is a double glazed window, which comprises two 6 mm thick acrylic sheets separated by an 18 mm gap filled with water. Such windows can sustain the unfocused laser beam for several minutes before burn through of the first sheet occurs. Moreover, if a sensitive water level sensor is incorporated, the moment water leaks from the window, if burn through of the first sheet occurs, the sensor can signal the laser to be switched off. The windows in the 5 axis laser gantry enclosure shown in Fig. 12.1 employ this principle.

- When setting up an Nd:YAG laser it may be necessary to approach the laser gun and make adjustments whilst using a low laser power. In such circumstances protective goggles, complying with the relevant national protection of eyes regulations (in the UK, Eyes Regulation, 1974, Certificate of Approval No. 2 (5)) must be worn. Moreover, the operator must position him or herself several lens focal lengths from the focused spot and in a position where there is no risk of reflected light from the work.

**12.1** The safety windows of this 5 axis $CO_2$ laser gantry enclosure employ double glazed acrylic sheets separated by a water filled gap; see text for details. The operator also watches the welding operation on a CCT monitor.

- When positioning workpieces for laser welding, it is particularly important to assess the likely direction of specular reflections from the workpiece. Particular attention should be paid to the inclination of the weld zone in relation to the position of any viewing windows, so that the viewing window does not lie in the path of any predictable specular reflections from the weld zone. This aspect requires special attention when working with high reflectivity materials such as aluminium and copper.
- Access to laser gun enclosures (workstations) should only be by authorised personnel who understand the potential laser beam hazards. The correct labels indicating the existence of a laser beam (behind a panel, etc) and the need to wear safety goggles should be displayed to act as a continual reminder.
- Skilled operators and supervisors must keep a close eye on junior staff and others under equipment operation and maintenance training.

### Electrical precautions

(The information presented here refers to the laser only.)

- Laser electrical systems fall into three general voltage levels: low voltage (typically 24 V); mains voltage (115–450 V) and high voltage (in the kV range). It is with the last two that the greatest care must be taken. However, the maintenance and repair of all these must only be undertaken by skilled electrical engineers and in accordance with the laser manufacturer's instructions where applicable.
- Before removing covers for service or repair one is advised always to isolate the equipment from the electrical supply. If possible padlock the isolator switch in the off position and if this is not possible remove the fuses. Also attach a notice to the isolator switch warning that work is in progress. Establish the position of all high voltage capacitors and discharge them to earth using the correct earthing probes (this action is advised even if the capacitors are fitted with self-dump circuits).
- When replacing parts always ensure that cables and terminal connections are correctly positioned to prevent the possibility of arcing between connections.
- For calibration and component set-up purposes it may be necessary to work on live circuits whilst carefully following the manufacturer's instructions (usually a special safety interlock override key switch is provided for these situations). In such circumstances the engineer should never work alone and the procedure for the work should ideally be discussed and agreed with the Laser Safety Officer. Moreover, always place notices warning other people in the vicinity that there is an electrical hazard and, where necessary, place no entry barriers around the danger zone.
- Clear labelling of electrical hazards, especially those to be found behind removable panels and access doors, must be displayed. If in any doubt about the correct form of labelling the company Safety Officer or the local health and safety inspector should be consulted.

Chemical and fume precautions

- In the event of suspected thermal damage to gallium arsenide and zinc selenide optical materials, which if overheated produce powdery deposits of toxic oxides, several minutes should be allowed for any oxides to settle before inspection takes place. For the inspection, a face mask and rubber, or strong polyethylene, gloves should be worn. If powdery deposits are present, or if in any doubt, the optic should be swabbed with wet tissues to pick up the oxide. The tissues should then be rinsed in several litres of water to disperse the concentration of the powdered oxide and the water disposed of in the waste water system.
- If on inspection, gallium arsenide and zinc selenide optical materials are found to be unserviceable and unsuitable for rework, they must be treated as toxic waste and disposed of accordingly.
- All welding fume should be regarded as potentially dangerous and therefore fume extraction systems should be used and the welding zone well ventilated (the ventilation should not disrupt the shielding gas above the weld keyhole). There are two basic types of extraction system: those that remove only particulates and those which remove particulates and gases. For welding applications, the latter type are usually an extension of the first, simply by directing the exhaust outside the building at a high level and away from windows, air intakes and occupied areas. The choice of type and size of extraction system will depend on the materials to be welded, the production volume and the concentration of fume produced. The UK Health and Safety Executive list of Occupational Exposure Limits, Guidance Note EH40, states that the concentration of welding fume must not exceed 5 $mg/m^3$ (this guidance note is revised annually and therefore the figure given here must be checked). Assessment of such an exposure level is not simple, requiring special equipment and test personnel. Consequently, one is advised to define the application clearly and consult a reputable fume extraction system supplier for a specification which will adequately fulfil service requirements.

## Further information on the safe use of lasers

Ignoring laser safety is as safe as jumping out of an aeroplane without a parachute. Safety is a moral obligation and this chapter has given an insight to perhaps the most pertinent aspects. However, for more detailed information in the UK and USA the following are suggested.

1 BS 7192:1989 Radiation safety of laser products - BSI Sales, Linford Wood, Milton Keynes, MK14 6LE, UK.
2 The Factories Act 1961 - HMSO, 49 High Holborn, London WC1V 6HB, UK.
3 The Protection of Eyes Regulations 1974 - HMSO.
4 Health & Safety at Work etc Act, 1974 - HMSO.
5 Laser Safety Guide - Laser Institute of America. Available from 5151 Monroe Street, Toledo, OH 43623, USA.
6 American national standard for the safe use of lasers (ANSI Z1316.1 - 1986). Laser Institute of America, 5151 Monroe Street, Toledo, OH 43623, USA.

Chapter 13

# Installing and operating a laser – some questions and answers

The information presented here is based on questions and answers, because most people having once read this book will have a basic knowledge of the technology and if contemplating installing and operating a laser at a later date, will probably refer directly to this chapter for answers to what, from experience, have proved to be common questions!

Laser welding equipment and indeed the associated technology is quite different from most other welding techniques. Chapters 2 to 4 indicate how laser welding differs with respect to the type of weld formation allowing a wide range of joint configurations to be welded, new component design opportunities and increased production rates. The subsequent chapters show that although laser welding may be tolerant to variations in laser power and welding speed, it is a precision process which requires attention to detail, plus special safety requirements, for its successful application. Despite these aspects, the chain of requirements for successful laser welding starts at the laser installation, the necessary supporting equipment and services and operating procedures.

The aim of this chapter is to introduce the basic considerations which are fundamental to the successful installation and operation of a welding laser and to give useful advice and to point out where related help can be found in this book. The aspects considered are: the laser; laser beam transmission and focusing equipment; gas shielding and equipment; work handling equipment; material preparation; laser safety; site preparation; achieving repeatable weld quality; operator and maintenance engineer training; and finally a brief outline for planning from concept to production.

The following information is based on the assumption that laser welding has been correctly chosen in the first place by a systematic evaluation, such as described in Chapter 2.

## The laser

- What type of laser and power is necessary to meet the required weld characteristics and welding speed?

The answer to this question will of course depend on the weld depth, weld shape and welding speed required to meet weld performance and production. From Chapters 1, 3 and 5 it can be seen that the present high power Nd:YAG lasers (1 kW) can confidently be considered for producing welds up to 4 mm deep at a speed of 0.3 m/min. For weld depths in the region of 1 mm they can achieve numerous weld shapes at substantially higher welding speeds, especially when operated in a micro-spot welding mode. However, for making deeper and continuous welds the welding speed and weld shape control diminish significantly as weld depth is increased, and if welding speed is a major requirement, then high power (1 kW) Nd:YAG lasers lose out to the more powerful $CO_2$ lasers. Nevertheless, for weld depths up to 4 mm, and where welding speed is not important, Nd:YAG lasers should be given serious consideration. This is because they are much more simple and robust than their $CO_2$ counterparts. Furthermore, their ability to employ fibre optics for beam transmission (Chapters 10 and 11) enables more simple, safer (the beam is enclosed in an armoured cable) and reliable transmissions of the laser beam to single or multiple workstations. Also the focusing optics for Nd:YAG lasers are much cheaper and the overall long term consumables are less. For a comparison of laser machine simplicity refer to Chapter 1 and compare the major component parts shown in Fig. 1.8 and 1.11. Nd:YAG lasers do, however, need special safety requirements with respect to eye protection, see Chapter 12.

The approximate laser power and necessary type for a given weld penetration depth and welding speed can be gauged from Chapter 5, which for $CO_2$ lasers, provides data for single pass welding material thicknesses up to 15 mm. If a laser is to be chosen for a specific welding application requiring a set weld depth and welding speed, it is recommended that a laser with at least 20% more power than is required is chosen. This will provide some flexibility for variations in parameters, but more important it will ensure the laser is not required to run at maximum performance and thus give no leeway if the laser output power should drop due to unforeseen circumstances; such as ageing of resonator optics, cathodes or flash lamps, etc.

Before purchasing a laser the supplier should be asked to produce welded samples which represent the product expected from the laser in service. It is recommended that the purchaser also witnesses the production of the samples, which should ideally be made consecutively and in a sufficient number to enable careful production and product evaluation.

- Will more than one laser be necessary to meet production demands?

One of the major advantages of a laser is its ability to work several workstations sequentially, and in special cases simultaneously, to achieve high production rates (Chapter 10). However, assuming the laser system can just meet production rates, if used for large volume production and 'just-in-time' delivery of welded components, it would be unwise to depend on one laser. Experience has indicated that well-maintained lasers generally achieve a serviceability of 85%. The remaining 15% is usually spread across servicing, repairing component failures, the delivery of spare parts and the travelling time of service engineers in the event of an unforeseen breakdown. Therefore, for large volume 'just-in-time' delivery, it would be an advantage to have more than one laser and also have spare workstation capacity, so that near or full production can be maintained if a laser becomes unserviceable. For further information see second question under heading 'Workhandling equipment'.

- What services are required to operate a laser?

The two major requirements for both $CO_2$ and Nd:YAG lasers are electricity and cooling water. To the newcomer, the required capacities of these are often considerably higher than expected. Although the output power from a laser can be used very efficiently when compared with electric arc techniques (Chapter 2), the total electrical input to output power conversion is poor, typically 3% for Nd:YAG lasers and 5% for $CO_2$ lasers (total machine system). Therefore, for every kilowatt of laser power, the input power required will be in the order of thirty-three times for an Nd:YAG laser and twenty times for a $CO_2$ laser. (Because of these poor conversion factors, for minimum weld costs, it is necessary to keep the total welding operation periods and speeds as high as possible in relation to the laser running times.)

The excitation of the laser medium causes considerable heat, which if not removed by water cooling, degrades photon generation (Chapter 1) and will cause damage to the laser resonator. Nd:YAG lasers generally require higher cooling water flow rates per kilowatt than $CO_2$

lasers. A typical 1 kW Nd:YAG laser will require approximately 50 l/min at a temperature in the order of 10–15 °C, whereas a 5 kW $CO_2$ laser will usually require between 65 and 80 l/min at a similar temperature. At these flow rates and temperature range, water from the public supply cannot be used and consequently a closed loop water chiller system is necessary. Water chillers for lasers are usually stand alone units, which are parked close to the laser. A laser input to output water temperature rise is typically in the order of 10 °C and therefore a chiller delivering chilled water at 50 l/min would be rated in the order of 35 kW and would require a floor space in the region of 1–1.5 $m^2$. Where a number of lasers are used in one area, floor space can be saved by using one or two large external chillers feeding a cooled water ring main. Large chillers can also have very effective waste heat recovery systems which feed the normally waste heat, in the form of hot air, back into the workshop in winter. Such a heating system is used in the laser research laboratory of TWI at Abington, England, see Fig. 13.1.

Apart from electrical and cooling water supplies some lasers also require compressed air for operating beam shutters and laser beam steering mirrors. $CO_2$ lasers require $CO_2$, $N_2$ and He supplies for producing the lasing action.

**13.1** A 300 kW water cooling plant, with waste heat recovery for heating the building (courtesy of Tri-Therm).

- What major spare parts should be held?

This is not an easy question to answer. When once asking a particular laser manufacturer the same question, the number of spare parts suggested would have warranted a small warehouse. From experience, the following parts are suggested for Nd:YAG and $CO_2$ lasers respectively.

*Nd:YAG laser*
Flash lamps
Resonator optics
Beam focusing optics
Focusing optic cover slides
Cooling water filters
Fuses
Shielding gas flow gauge and regulator
Resonator cavity O ring seals

*$CO_2$ laser*
Resonator optics
Beam transmission optics
Beam focusing optics
Cathodes (a complete exchange set for a DC excited laser)
Ballast resistors
Fuses
Laser cavity gas cooling heat exchanger
Vacuum pump oil seals, or spare pump if laser is used continuously
Vacuum pump oil
Blower oil seals, or exchange blower if laser is used continuously
Blower oil
Cooling water filters
Shielding gas flow gauge and regulator
Resonator cavity O ring seals
Aero-window pump oil filters and oil (if fitted)

## Laser beam transmission and focusing equipment

- What type of laser beam focusing optic and transmission equipment is necessary?

These will depend on the type of laser, the power required and the distance of the laser from the workstation, or stations. Taking the beam focusing optic first, for an Nd:YAG laser, transmissive glass optics

(Chapter 11) are the norm and often in a form which comprises a beam expansion and or collimating lens preceding a focusing lens. For $CO_2$ lasers, zinc selenide and potassium chloride transmissive optics can be used for laser powers up to 5 kW, but where possible and for higher laser powers, gold coated copper, copper and molybdenum reflective optics (Chapter 11) are recommended as they are more durable.

The focal length of the beam focusing optic in relation to its incident beam diameter (the f number) will need to be given serious consideration as it determines the focus spot size (controls the welding power density) and the tolerances for focus position (Chapter 5) and laser beam joint alignment (Chapters 6 and 7). When purchasing a turn-key laser system for a given welding application, the system supplier will normally take care of this aspect, however, if this is not the case an introduction and practical guide is given in Chapter 5.

The laser beam focusing optic is prone to damage by weld spatter and metal vapour and therefore serious consideration should be given to employing an optic protection device (Chapter 11).

Where a laser is to be operated in conjunction with a small single workstation, it may not be necessary to use a purpose-built beam transmission system as the laser gun can often be fixed directly on the end of the laser with the aid of a short connection tube and beam fold mirror assembly to direct the beam down into the laser gun (Chapter 11). Where the workstation cannot be directly aligned with the laser output window or multiple workstations are required, a fibre optic and beam switching system will be necessary for an Nd:YAG laser and an enclosed water cooled mirror train and beam switching system for a $CO_2$ laser (Chapters 10, 11 and 12).

## Gas shielding and equipment

- What type of shielding gas, device and control is required?

When making single spot welds with very short weld pulse lengths a gas shield may not be necessary. However, in all other cases, especially when making continuous welds, a shielding gas is most important. In the main, helium is used with $CO_2$ lasers, and argon with Nd:YAG lasers, and in certain cases nitrogen can be used with both laser types. Information on these and other possible gas mixtures is presented in Chapter 5.

There are numerous gas shield dispensing devices and the choice will depend on the joint configuration, joint access and whether the laser gun is manipulated about the work or vice versa. Various designs are described and illustrated in Chapter 5.

The necessary flow rate for the shielding gas depends on the laser power (Chapter 5) and the rate increases with laser power. Variations in shielding gas flow rate can have a major effect on weld formation and therefore careful control and monitoring is necessary (Chapter 9).

## Workhandling equipment

- What type of workhandling equipment is necessary?

This requirement will depend on the component shape and size. If the component is large or heavy it may be more practical to manipulate the laser gun about the workpiece. However, apart from instances where a simple linear gun movement is required, such an approach should always be a last resort, because multi-axis gun manipulation systems, especially for $CO_2$ laser guns, are expensive and difficult to maintain. Circumferential tubular butt, lap seam, annular butt and T butt welds, where the weld line remains in one plain, can usually be made with conventional rotary welding tables. Other manipulatable components may need linear and programmable X-Y tables in conjunction with vertical or various other laser gun movements. Such concepts for workhandling and beam manipulation are described in Chapter 10, which should give an insight to the best approach for welding the components in question.

- Will laser time sharing welding stations be necessary and if so how many?

Whether time sharing stations are necessary will depend on the required daily production rate in relation to the most efficient workhandling and welding cycle time which can be achieved on a single station. If the daily production required exceeds the single station production, then the number of stations required will equal the production required divided by the single station daily output. The resultant number of stations may of course exceed the number which can be handled by one laser. To establish the number of stations which can be handled, divide the single station workhandling (work loading and unloading) time by the weld time.

- What are the most common factors which are overlooked when installing workhandling and laser time sharing stations?

The most common factor is perhaps adequate floor space to enable safe access for servicing (especially safe laser beam alignment) and positioning of safety screening. Equipments are often squeezed into production lines and therefore these aspects often suffer. Other aspects often overlooked are the necessary spares, service intervals and procedures.

## Material preparation

- Why is material preparation so important?

Laser welds are very narrow when compared with most other fusion welding techniques (Chapter 2) and therefore, except when making lap seam joints, precision alignment of the focused laser beam with the joint line is required (Chapters 6 and 7). Consequently, the material faces to be joined must be prepared in a manner which will provide close fitting joints (that is unless a filler material is being used, Chapter 8).

Laser welding does not use fluxes to clean the weld faces and remove contaminates from the weld pool, so a part cleaning operation is usually necessary just prior to the welding operation especially where high quality welds are required (Chapter 9).

- If chemical cleaning is used will fume extraction be necessary?

Most cleaning agents, particularly degreasing agents, give off vapours which can affect the transmission of the laser beam (Chapter 11) and therefore, unless the parts can be cleaned in a ventilated area well away from the laser system, fume extraction is strongly recommended. This is not only for laser operation, but also for the protection of the operator and other personnel (Chapter 12).

Always use clean cleaning agents, otherwise they can leave behind more fine particulate contamination than they remove and thus degrade the weld quality.

## Laser safety

- What are the most important points to be aware of with respect to laser safety?

All persons and operators in the vicinity of a welding laser must be fully aware of the potential hazards posed by laser light, the laser and the laser welding operation. Chapter 12 describes the potential hazards and necessary steps to safeguard against them. Besides this information, users are advised to consult their local health and safety inspector and if possible in conjunction with the laser or laser system supplier. A Laser Safety Officer should be appointed, as ignoring laser safety is extremely dangerous.

## Site preparation

- What are the major considerations when selecting and preparing the site for a laser welding system?

First one must establish if the site is suitable for operating a laser, i.e. that it is free from vibrations and floor movements, airborne vapours, humidity and sudden temperature changes, all of which affect running the laser and the welding operation. Vibrations and floor movements such as those produced by blanking machines and forming presses can seriously upset the optical alignment of both laser and the beam transmission systems, especially in the case of $CO_2$ lasers where mirror alignment is paramount for repeatable welding performance. If a $CO_2$ laser has to be operated where vibrations are transmitted through the floor, then the floor will need to be modified to provide an area for the laser which is vibration-free. Likewise, if the workshop floor is prone to movement, steps need to be taken to provide a movement-free area. The same rule also applies for the optic support pillars for the beam transmission system. Because Nd:YAG lasers are much smaller than $CO_2$ lasers they can usually be placed on a relatively small mount capable of damping serious vibrations and floor movements. Moreover, because the beam can be transmitted through an optical fibre, damping of any support pillars, other than that which supports the laser gun, is unnecessary.

Airborne vapours such as those from cleaning agents and cutting oil, and also high humidity, must be protected against, otherwise they will affect weld formation, beam transmission and optic life (Chapters 5 and 11).

If the environment is suitable, or can be made suitable, for operating a laser, then the next consideration is the floor area. Here one must ensure that the area is adequate for the total system, that is, the laser beam transmission system, workstation(s), the electrical, cooling water and gas supplies and if necessary part cleaning, and assembly equipment. Added to this and very important is an adequate area for servicing and repair; some $CO_2$ lasers require a substantial adjacent area for their disassembly.

Because lasers and their associated equipment can take up a large area and require environmental conditions which may be difficult to achieve, serious consideration should be given to the possibility of siting the laser away from the production area and piping the beam to the workstations. If preparing a 'green field site' it would be wise to consider having a purpose-built laser room near the production area. This would provide for easier maintenance, the storage of associated equipment and spare parts, and of course enhance safety.

## Achieving repeatable weld quality

- What are the most important aspects for achieving repeatable weld quality?

As for all welding techniques, achieving repeatable weld quality requires an unbroken chain of correct actions and events. For laser welding these are many and can be grouped under three major headings: materials, welding conditions and equipment operation and performance; all of which must be carefully established and maintained. Chapter 4 gives a guide to material selection and where to seek professional advice. Information on establishing welding conditions and associated requirements for sheet metal joints, thicker section joints and the use of filler material, is presented in Chapters 5 to 8. Once materials and welding conditions have been well established, care is then necessary to ensure that they stay that way by applying quality assurance techniques (Chapter 9). The same applies to the performance of the welding equipment and in this respect equipment monitoring techniques can play an important part.

- Which monitoring equipments are necessary to achieve a reliable welding operation?

There are numerous monitors available for setting up the welding conditions, examining the laser performance and indicating weld formation (Chapter 9). Some monitors are essential whereas others are useful diagnostic tools for spotting in-process system malfunctions and troubleshooting. For setting welding conditions the two most essential monitors are an absorption type laser power measuring probe, which can measure the unfocused laser beam power near the position of the focusing optic, and a device for checking and measuring the welding speed. For the welding operation, flow meters fitted with devices for sensing the correct flow of weld shielding gas to the work and cooling water to the laser beam transmission optics are recommended.

In-process laser beam and weld quality monitors (Chapter 9) need expert selection and interpretation of their signals. They are perhaps best acquired only after careful trials to prove their suitability.

- Which is the best method of assessing welding performance during production?

There is no single answer to this question because the method of assessment will depend on the weld service requirement and the ease with which the necessary tests can be carried out to ensure the service requirements are met. Sometimes every weld can easily be proof tested and subsequently put into service with confidence. Other welds, such as those used for lap seam joints between thin sheets, can be visually appraised with confidence by observing that the sheets show no separation, the weld has fully penetrated the lower sheet and shows no signs of defects such as blowholes or cracks. Unfortunately, more often than not, most welded joints can only be assessed by destructive testing, partial penetration welds joining gears for example, necessitating the removal of samples at regular intervals from the production line for testing. There are, however, sometimes instances where non-destructive monitoring techniques such as ultrasonic testing can be used. These and other techniques are described in Chapter 9.

## Operator and maintenance engineer training

- What personnel abilities are necessary to operate and maintain a laser?

Successful laser welding requires attention to detail, as it is a precision process in terms of machine set-up and operation. Therefore, unless the welding station is a turn-key type and maintained by a trained engineer, the operator needs to be adept, especially with respect to setting welding conditions. The operator must also be capable of clearly understanding the necessary laser safety requirements. Generally speaking, machine shop personnel such as lathe and mill operators, who are used to making precision adjustments, make good laser operators, enjoy the challenge and become 'champions' with respect to operation and productivity.

Laser maintenance can be carried out by the supplier, but this can prove impractical (due to travelling and waiting time) and expensive for routine servicing such as optic and flash lamp replacement, optic and cathode cleaning and the maintenance of water pumps, filters, vacuum pumps, blowers, etc. Consequently, an engineer trained in laser maintenance is recommended. If a person has to be selected for training they ideally should have a proven mechanical and or electrical engineering background and above all a sound knowledge of the potential hazards when working with high voltage electrical equipment (Chapter 12).

- Where can operator and maintenance training be obtained?

Most of the major laser machine and equipment suppliers offer training facilities, and it is wise to establish if this requirement is available before purchasing a laser or laser system. Professional organisations such as the TWI Laser Centre, in the UK and The Laser Institute of America, in the USA, will provide teach-ins and technical information on operating lasers and laser safety.

- Once the required personnel are trained, how does one ensure the necessary operation and maintenance procedures are maintained?

Laser operation and maintenance procedures should be prepared, if not supplied by the machine vendor, and kept in a handy place. The personnel concerned should regularly refer to these documents to refresh their memories, also their supervisor should ensure this action takes place, especially with respect to the laser safety aspects (Chapter 12).

## Planning from concept to production

All the aspects discussed in this chapter contribute to installing and operating a laser welding system. Those who have already installed a laser will no doubt be able to add numerous other details for consideration. Others reading this chapter may well be starting out for the first time. To assist these people, and those who have a general interest, Fig. 13.2 provides a typical programme for taking laser welding from conception to production. It also perhaps provides a summary of many of the aspects which are covered in this book.

## Epilogue

The next chapter is a glossary of laser and welding terminology and therefore this chapter really concludes, from a practical know-how

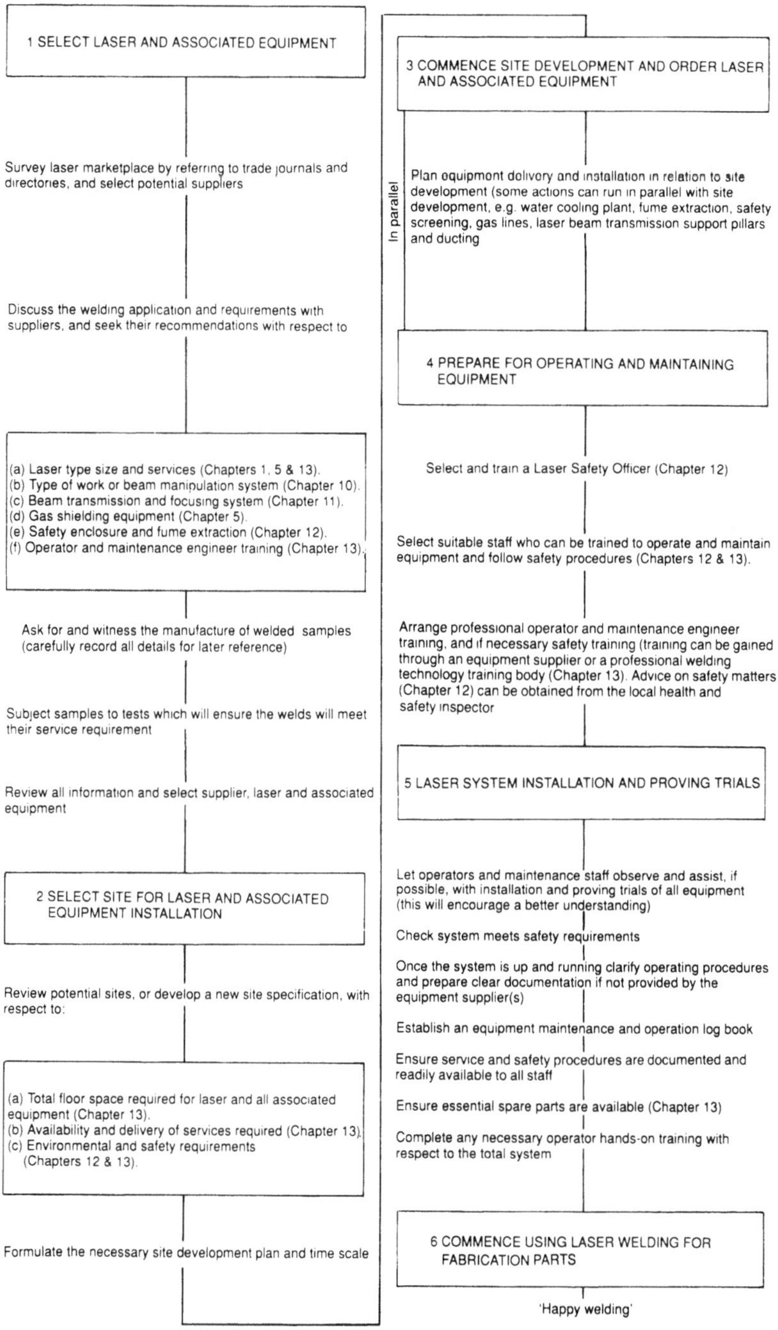


**13.2** Installing a laser – planning from concept to production.

standpoint, this practical guide to laser welding. It is hoped, for those to whom this book is an initial introduction to laser welding, that they have gained an insight to a welding technique that has given the author many hours of pleasure and fascinating challenge with respect to its application and development. The aim has been to help people to get started with information which is proven, but not claimed to be optimum, and to provide a source for future reference.

One final point, although laser welding offers design and production opportunities which may not be available from other welding techniques, it is by no means the be-all and end-all of welding technology. It must be remembered that every welding technique has its advantages and disadvantages. Always take care to choose the correct welding technique in the first place!

Chapter 14

# Glossary of common terminology

**Absorb**
To transform radiant energy into a different form, usually with a resultant rise in temperature.

**Absorbance**
The ability of a medium to absorb radiation, depending on temperature and wave length. Expressed as the negative common logarithm of the transmittance.

**Active medium**
A medium in which lasing will take place, rather than absorption, at a given wave length.

**Afocal**
Literally, 'without a focal length'; an optical system with its object and image point at infinity.

**Amplification**
The growth of the radiation field in the laser resonator cavity. As the light wave bounces back and forth between the cavity mirrors, it is amplified by stimulated emission on each pass through the active medium.

**Amplitude**
The maximum value of the electromagnetic wave, measured from the mean to the extreme; i.e. the height of the wave.

**Angle of incidence**
See **incident ray**.

**Angstrom unit**
A unit of measurement for a wave length of light (written A), equal to one ten billionth of a metre ($10^{-10}$ metre). Occasionally used.

**Anode**
An electrical element in laser excitation which attracts electrons from a cathode. An anode can be cooled directly by water or by radiation.

**AR coatings**
Anti-reflection coatings, used on the backs of laser output mirrors and lenses to suppress unwanted multiple reflections which reduce power.

**Austenite, austenitic**
Gamma iron with carbon in solution; work-hardening stainless steels that cannot be hardened by heat-treating. Named after metallurgist Sir W C Roberts-Austen.

**Autocollimator**
A single instrument combining the functions of a telescope and a collimator to detect small angular displacements of a mirror by means of its own collimated light.

**Autogenous welding**
Fusion welding without the addition of filler metal.

**Axial flow laser**
The simplest and most efficient of the gas lasers. An axial flow of gas is maintained through the tube to replace those gas molecules depleted by the electrical discharge used to excite the gas molecules to the lasing state, see **gas discharge laser**.

**Axis, optical axis**
The optical centreline for a lens system; the line passing through the centres of curvature of the optical surfaces of a lens.

**Beam bender**
An optical device, such as a mirror, capable of changing laser beam direction; used to repoint the beam and in 'folded', compact delivery systems.

**Beam diameter**
The diameter of that portion of the beam which contains 86% of the output power.

**Beam expander**
Optical device increasing beam diameter and reducing divergence. Result: A smaller focused spot for more distance between lens and part.

**Beam splitting**
Optical splitting of a laser beam into two or more beams, allowing work on more than one side of a part at the same time but at somewhat less power than with a multiple-output beam system.

**Bend test**
A test for ductility, usually applied by bending material over a former of specified radius. There are national and international standards for weld bend tests.

**Black body**
One which totally absorbs incident radiation, or which, when incandescent, emits a continuous spectrum.

**Blank**
A metal sheet prepared for some subsequent forming operation, such as stamping, piercing or extrusion.

**Blanking**
Cutting shapes from metal sheets.

**Blowholes**
Casting and weld metal defects, often elongated and with smooth walls. Caused by air or gases trapped in the metal as it solidifies.

**Brewster windows**
The transmissive end (or both ends) of the laser tube, made of transparent optical material and set at Brewster's angle in gas lasers to achieve zero reflective loss of vertically polarized light. Non-standard on industrial lasers, but a necessity if polarization is desired.

**Brightness**
The visual sensation of the luminous power of a light beam, as opposed to scientifically measured power of the beam.

**Brittle fracture**
A failure mechanism which can occur in structural steels, both plain carbon and those of low alloy content, when subjected to low temperatures.

**Calorimeter**
An instrument which measures the heat generated by absorption of the laser beam; another way to measure laser power.

**Carbide**
A combination of carbon with one or more other elements.

**Carbon**
A non-metallic element, important in ferrous metals for introducing hardness and strength by the formation of carbides.

**Carbon dioxide**
$CO_2$, the product of complete combustion of carbon.

**Carbon equivalent**
An empirical relationship used to determine the equivalent carbon content of low alloy or high manganese carbon steel for weldability purposes. See Chapter 4 for further details.

**Cathode**
The element providing the electrons for the electrical discharge used to excite the lasing medium.

**Cathodic protection**
Reduction in the corrosion rate of the metal to be protected by making its potential more negative. Achieved by using a sacrificial anode material such as an electroplated or hot-dipped zinc coating.

**$CO_2$ laser**
A laser largely used in industry in which the primary lasing medium is carbon dioxide.

**Coaxial gas**
Most laser welding is carried out with the shield of inert gas flowing over the work surface to prevent plasma oxidation and absorption, to blow away debris and to control heat reaction. The gas jet has the same axis as the beam so the two can be aimed together.

**Coherent light, coherent radiation**
Radiation composed of wave trains vibrating in phase with each other. Simply expressed: parallel rays of light.

**Collimated light**
Divergent light rays rendered parallel by means of a lens or other device, allowing a sharp image of the object to be focused at the focal plane of the lens.

**Collimation**
The process by which divergent rays (white, or natural light) are converted into parallel rays (coherent light).

**Continuous wave**
CW, the continuous emission mode of a laser opposed to pulsed operation.

**Convergence**
The bending of light rays toward each other, as by a positive (convex) lens.

**Corrected lens**
A compound lens that is measurably free of aberrations through careful selection of dimensions.

**Corrosion**
Chemical or electrochemical reaction between a metal and its environment, usually resulting in deterioration in the properties and/or the appearance of the metal.

**Corrosion resistance**
A vague term indicating the general corrosion resistance of a material.

**Current regulation**
Laser system regulation in which discharge current is kept constant at all times, regardless of output power (which can change because of optical misalignment or dirt).

**Current saturation**
Maximum flow of electric force in a conductor; in a laser, the point at which further electrical charge will not increase lasing action.

**Deep drawing**
Drawing a cup shape of large depth to width ratio from sheet or strip material.

**Depth of field**
The working range of the beam, a function of wave length, diameter of the unfocused beam and focal length of the lens. To achieve a small diameter spot size, and thus a high power density, a short depth of field must be accepted.

**Divergence**
The angle at which the laser beam spreads in the far field; the bending of rays away from each other, as by a concave lens or convex mirror.

**Downhand welding**
Welding along a horizontal line and surface.

**Drift, angular drift**
All undesirable variations in output (either amplitude or frequency); angular drift of the beam, measured in milliradians before, during and after warm-up.

**Duty cycle**
The length of time the laser beam is actually welding, as compared with the entire work cycle time.

**Electric vector**
The electric field associated with a lightwave and having both direction and amplitude. Commonly represented by a line with an arrowhead.

**Electromagnetic wave**
A disturbance which propagates outward from an electric charge which oscillates or is accelerated. Includes radio waves, X-rays, gamma rays and infrared, ultraviolet and visible light.

**Emissivity emittance**
Rate at which emission takes place; the ratio of the radiant energy emitted by a source or surface to that emitted by a black body at the same temperature.

**Enhanced pulsing**
Electronic modulation of a laser beam to produce high peak power at the initial stage of the pulse. This allows rapid vaporization of the material without the surrounding area.

**Erichsen test**
A cupping test to establish the suitability of a material for deep drawing. A 20 mm diameter ball is forced into a sheet of material via the centre of a pair of flat rings used to clamp the sheet lightly. With the aid of the ball a dome is forced into the sheet and its height before cracking indicates the sheet material suitability for pressing.

**Exposure**
A measure of the total radiant energy incident on a surface per unit area; radiant exposure.

**Far field imaging**
An imaging technique with solid-state lasers that has several limitations: non-uniform energy distribution, very short working distances, poor control of hole geometry.

**Fatigue**
Failure of a material by repeated application of stresses, each of which is individually insufficient to cause failure.

**Ferrite**
A body-centred cubic (alpha) form of iron which is stable below 760°C in the pure form. In steels, many elements enter into solid solution in the ferrite, so the term is applied to any solid solution based on alpha iron.

**Ferritic steels**
Steels in which ferrite is the predominant phase.

**Flashlamp**
Source of powerful light; often in the form of a helical coil and used to excite photon emission in a solid-state laser.

**Flux**
The radiant, or luminous, power of a light beam; the time rate of the flow of radiant energy across a given surface.

**Focus**
As a noun, the point where rays of light, which have been reflected by a mirror or refracted by a lens's meet, giving rise to an image of the source. As a verb, to adjust focal length for the clearest image.

**Focal point**
Same as first definition under **focus** in laser work, the focal point of the beam relative to the work surface has a critical effect; it controls the power density.

**Folded resonator**
Construction in which the interior optical path is bent by mirrors mounted on corner blocks bolted into pre-aligned positions, permitting compact packaging of a long laser cavity.

**Frequency**
The number of light waves passing a fixed point in a given unit of time, or the number of complete vibrations in that period of time.

**Fully killed steel**
A steel which has been fully deoxidised.

**Fusion welding**
Any welding operation which involves melting of the parent metal.

**Gain**
Another term for amplification, usually referring to the efficiency of a lasing medium in attaining a population inversion. High gain is typically more than 50% per pass of the light wave between cavity mirrors.

**Gas discharge laser**
A laser containing a gaseous lasing medium in a glass tube in which a constant flow of gas replenishes the molecules depleted by the electricity or chemicals used for excitation. The discharged gas can be filtered and 90% recycled for economy.

**Gas transport laser**
A laser design which generates very high beam power within a fairly small resonator structure. Long electrodes parallel the axis and gas is circulated *across* the resonator cavity. See Chapter 1, cross flow laser.

**Gaussian**

The 'normal curve' or normal distribution, an example of which is the symmetrical bell shape of the holes created by the uncorrected, unfocused laser beam in its optimum mode. A Gaussian laser beam has most of its energy in the centre.

**Heat affected zone**

HAZ, the zone within the parent metal of a welded joint in which heat has caused structural changes without melting.

**Helium-neon laser**

A laser in which the active medium is a mixture of helium and neon, which is on the visible range. Used widely in industry for laser alignment operations, also for recording, printing and measuring.

**Heat sink**

A substance or device used to dissipate or absorb unwanted heat, as from a manufacturing process (or, with lasers, from reflected rays).

**Heat treatment**

A thermal treatment applied to a metal to change its structure.

**Hydrogen embrittlement/cracking**

Loss of ductility or toughness in a metal as a result of the presence of hydrogen within the material. Can cause delayed fracture.

**Incident ray**

A ray of light that falls on the surface of a lens or any other object. The 'angle of incidence' is the angle made by the ray with a line perpendicular to the surface.

**Ion laser**

A type of laser employing a very high discharge current, passing down a small bore to ionize a noble gas such as argon or krypton. The ionization process creates a population inversion for lasing to occur. A research laser useful for some industrial applications.

**Intensity**

The magnitude of radiant energy (light) per unit, such as time or reflecting surface.

**Irradiation**

Exposure to radiant energy, such as heat, X-rays, or light; the product of irradiance and time.

**Joint efficiency**

The ratio of the strength of a joint as compared with the strength of the parent metal. Usually measured in tension.

**Joule**

One watt per second; a measurement frequently given for laser output in pulsed operation.

**Keyholing**

In welding, the deep-penetration holes, which fill quickly with molten metal, that can be made in a few milliseconds by laser.

**Laser**

An acronym of *L*ight *A*mplification by *S*timulated *E*mission of *R*adiation. A laser is a cavity, with mirrors at the ends, filled with lasable material such as crystal, glass, liquid, gas or dye. The materials must have atoms, ions or molecules capable of being excited to a metastable state by light, electric discharge or other stimulus. The transition from this metastable state back to the normal ground state is accompanied by the emission of photons which form a coherent beam. See Chapter 1.

**Laser oscillation**

The build-up of the coherent wave between laser cavity end mirrors. In CW mode, the wave bounding back and forth between mirrors transmits a fraction of its energy on each trip; in pulsed operation, emission happens instantaneously.

**Laser rod**

A solid-state, rod-shaped lasing medium in which ion excitation is caused by a source of intense light, such as a flashlamp. See Chapter 1.

**Laser texturizing**

Metals can be laser-textured by rapid vaporization creating a pitted surface without distortion of the product, e.g. dimensional tolerance is 5 to 7 times better than with sandblasting.

**Leading edge spike**

The initial pulse in a series of pulsed laser emissions, often useful in starting a reaction at the target surface. The trailing edge of the laser power is used to maintain the reaction after the initial burst of energy.

**Light**

The range of electromagnetic radiation frequencies detected by the eye, or the wavelength range from about 400 to 750 nanometers. It is sometimes extended to include photovoltaic effects and radiation beyond visible limits.

**Light regulation**
A form of power regulation in which output power is maintained at a constant level by controlling discharge current.

**Luminance**
Commonly called illumination; the luminous or visible flux per unit area on a receiving surface at any given point.

**Mandrel**
An axle, spindle, or other device that holds or supports a piece of work for machining.

**Martensite, martensitic**
The hard constituent of which quenched steel is chiefly composed; hardenable steel.

**Meniscus lens**
A lens which has one side convex and the other concave, see Chapter 11.

**Metastable, metastable state**
Unstable condition in which the energy of a molecule is at some discrete level above the lowest, or ground state. It is this condition which is necessary for emission of photons in a laser (from quantum theory).

**Mild steel**
A vague generic term for carbon steels containing 0.1–0.25% carbon.

**Mode**
A particular functioning arrangement, set-up, or condition for laser operation, such as continuous emission, pulses, or grouped pulses. 'Mode' also describes the cross-sectional shape of the beam, as described in Chapter 1.

**Modulation**
The ability to superimpose an external signal on the output of the laser as a control.

**Monochromatic light**
Theoretically, light consisting of just one wave length. As no light is completely monochromatic, it usually consists of a very narrow band of wave lengths. Lasers provide the narrowest bands.

**Near field imaging**
A solid-state laser imaging technique offering control of spot size and hole geometry, adjustable working distance, uniform energy distribution and easily produced range of spot sizes.

**Nd:glass laser**
A solid-state laser of neodymium:glass offering high power or short pulses, or both, for specific industrial applications.

**Nd:YAG laser**
A solid-state laser of neodymium:yttrium-aluminium-garnet, similar to the Nd:glass laser. Both are pumped by flashlamp.

**Nital**
A chemical etchant for steels comprising 2% nitric acid in alcohol.

**Non-destructive testing**
NDT, inspection without destruction of the sample or interference with its subsequent use for its normal purpose.

**Normalising**
A heat treatment used to soften a metal and also for relieving residual stresses arising from cold working or welding operations.

**Optical fibre**
Filament of quartz or other optical material capable of transmitting light along its conformation and emitting it at the end.

**Optical pumping**
Exciting the lasing medium by the application of light rather than electrical discharge from anode and cathode.

**Optical resonator**
See **resonator**.

**Output coupler**
The resonator mirror which transmits light; the one at the opposite end is totally reflective.

**Output power**
The energy per second emitted from the laser in the form of coherent light, usually measured in watts for continuous-wave operation and joules for pulsed operation.

**Pearlite**
The lamellar carbide-containing constituent present in annealed and normalised carbon steels and in grey cast iron, and characterised by a 'finger-print' like appearance of the alternate plates of iron and iron-carbide in the microstructure.

**Penetration**
The depth by which the fused zone of a weld penetrates the parent metal or previous weld deposits.

**Photometer**
An instrument which measures luminous intensity.

**Photon**
In quantum theory, the elemental unit of light, having both wave and particle behaviour. It has motion, but no mass or charge.

**Plasma**
In laser welding, a highly ionised gas or metal vapour that forms above the spot where the beam reacts with the metal surface. Also used to describe the laser tube (plasma tube, discharge tube) which contains the completely ionised gas in certain lasers.

**Pointing errors**
Beam movement and divergence, often preventable by using short path lengths.

**Polarization**
Restriction of the vibrations of the electromagnetic field to a single plane, rather than the innumerable planes rotating about the vector axis. This prevents optical losses at interfaces between the lasing medium and optical elements.

**Population inversion**
When more molecules (atoms, ions) in a laser are in a metastable state than in the ground state (a situation needed for sustaining a high rate of stimulated emissions), a population inversion is said to exist. Without a population inversion, there can be no lasing action.

**Power density**
The amount of radiant energy concentrated on a surface.

**Power meter**
An accessory used to monitor laser beam power.

**Power ramping**
A controlled change in the power of a laser beam, either linearly, as up a ramp, across and down again or in several discrete steps. Useful for smooth completion of circular welds and for preventing fractures from rapid cooling.

**Pulse energy**
The power of a single, brief emission from a laser programmed for pulsed behaviour rather than continuous operation. Pulse energy can be several times greater than CW emission.

**Pulse tail**
Pulse decay time, which can be shortened (by using a special mixture of gases) to allow for fast repetition of laser pulses within a given length of time.

**Pump**
To excite the lasing medium, see **optical pumping**.

**Q**
The energy-storing efficiency of a laser resonator. The higher the 'Q' the less energy loss.

**Q-switch**
A device that has the effect of a shutter moving rapidly in and out of the beam to 'spoil' the resonator's normal Q, keeping it low to prevent lasing action until a high level of energy is stored. Result: a giant pulse of power when normal Q is restored.

**Radian**
An arc in a circle, equal in length to the radius; an angle (57.3°) at the centre of a circle, formed by two radii cutting off such an arc. Thus one rad = 57.3°.

**Radiance**
Brightness; the radiant energy per unit solid angle and per unit projected area of a radiating surface.

**Radiant energy**
Energy travelling as wave motion; specifically, the energy of electromagnetic waves (light, X-rays, radio, gamma rays).

**Reflectance**
The ratio of the reflected flux to the incident flux, or the ratio of reflected light to light falling on the object.

**Reflection**
The return of radiant energy (incident-light) by a surface, with no change in wave length.

**Refraction**
The change of direction of propagation of any wave, such as an electromagnetic wave, when it passes from one medium to another in which the wave velocity is different. Simply put, the bending of incident rays as they pass from one medium to another, such as air to water.

**Refractory**
A term used to describe materials of high melting point.

**Residual stress**
A self-supporting internal stress system within a nominally unloaded component.

**Resolution**
Resolving power, or the quantitative measure of the ability of an optical instrument to produce separable images of different points on an object; the capability of making distinguishable the individual parts of an object, closely adjacent images, or sources of light.

**Resonator**

The mirrors (or reflectors) making up the laser cavity containing the laser rod or tube. The mirrors reflect light back and forth to build up amplification under an external stimulus. Emission is through one of them, called a coupler, which is partially transmissive.

**Rimming steel**

A steel which is only partially deoxidized before casting into the ingot.

**Root run**

The first weld at the narrowest section of a multi-pass weld.

**Shear strength**

The maximum shear stress which a material can stand.

**Spectral response**

The response of a device or material to monochromatic light as a function of wave length.

**Specular reflection**

Perfect or regular reflection of electromagnetic radiation, e.g. light. It occurs whenever a reflecting surface is flat to approximately 1/8 of a wavelength of the radiation incident upon it.

**Stainless steel**

A term used to describe a wide range of steels in which chromium is a common feature to enhance corrosion resistance, see Chapter 4.

**Stimulated emission**

When an atom, ion or molecule capable of lasing is excited to a higher energy level by an electric charge or other means, it will spontaneously emit a photon as it decays to the normal ground state. If that photon passes near another atom of the same frequency which is also some metastable energy level, the second atom will be stimulated to emit a photon. Both photons will be of the same wave length, phase and spatial coherence. Light amplified in this manner is intense, coherent (collimated or parallel) and monochromatic, in short, laser light, see Chapter 1.

**Stress**

Stress is a force per unit area. In metals, it may also be defined as the intensity of the force which resists a change in shape.

**Stress relieving**

A low temperature heat treatment to reduce the level of internal stress in a material.

**Surface tension**
The 'skin-like' tension at the surface of a liquid due to unbalanced molecular force there.

**Tack welds**
Small scattered welds made to hold a joint in alignment for welding.

**Tensile test**
Possibly the most widely used single mechnical test for material or weld strength. A specially prepared specimen pulled in the direction of its length until fracture occurs, the force to failure observed and the stress calculated.

**Toughness**
A general term implying the ability of a material to absorb energy by plastic deformation.

**Transmission**
In optics, the passage of radiant energy (light) through a medium.

**Transmittance**
The ratio of transmitted radiant energy to incident radiant energy, or the fraction of light that passes through a medium.

**Wave**
An undulation or vibration, a form of movement by which all radiant energy of the electromagnetic spectrum is thought to travel.

**Wave length**
The fundamental property of light, the length of light wave, which determines its colour. Common units of measurement (which is usually from crest to crest) are the micron, the nanometer and (earlier) the angstrom.

**Window**
A transparent material with plane parallel sides which admits light into or through an optical system and excludes dirt and moisture.

# Index

FSC
www.fsc.org
MIX
Papier | Fördert
gute Waldnutzung
FSC® C083411

Kolibri 360 GmbH
Ferdinand-Jühlke-Straße 7
99095 Erfurt, Deutschland
produktsicherheit@kolibri360.de